Food Consumer Science

Dominique Barjolle • Matthew Gorton
Jasna Milošević Đorđević • Žaklina Stojanović
Editors

Food Consumer Science

Theories, Methods and Application to the Western Balkans

Springer

Editors
Dominique Barjolle
Institute for Environmental Decisions
ETH Zürich, Zürich,
Switzerland

Matthew Gorton
Newcastle University Business School
Newcastle upon Tyne
Nairnshire, UK

Jasna Milošević Đorđević
Ipsos Strategic Marketing
Belgrade, Serbia

Žaklina Stojanović
University of Belgrade
Belgrade, Serbia

ISBN 978-94-007-5945-9 ISBN 978-94-007-5946-6 (eBook)
DOI 10.1007/978-94-007-5946-6
Springer Dordrecht Heidelberg New York London

Library of Congress Control Number: 2013934347

Printed on acid-free paper

Springer is part of Springer Science+Business Media (www.springer.com)

Foreword: Food Consumer Science at Crossroads

Wim Verbeke, Ph.D.

W. Verbeke
Full Professor in Agro-food Marketing and Consumer Behaviour, Department of Agricultural Economics, Ghent University, Coupure links 653, B-9000, Gent, Belgium
e-mail: wim.verbeke@ugent.be

Keywords Food consumer science • Health claims • Organic products • Traditional foods

Abbreviations

EU	European Union
EUFIC	European Food Information Council
TSG	Traditional Speciality Guaranteed

Food consumer science is at the crossroad of a multitude of scientific disciplines and interests. Research methods mainly originate from social sciences such as economics, marketing, psychology and sociology, while the most common applications bend towards the natural sciences related to agriculture, food and nutrition. The exploration of this crossroad presents a challenging and rewarding domain with numerous opportunities to contribute to a better understanding of (the determinants of) food choice and consumer decision-making and in extension also to more effective public policy, industry competitiveness, societal welfare and human well-being. Not only has the European Union (EU) undergone fundamental changes during the last 15 years, also food consumer science has been a field of continuous evolution. After a major focus on food safety, risk perception and risk communication since the mid-1990s, the attention has gradually expanded to cover also nutrition, diet and health and further to sustainability and food authenticity most recently. The number of peer-reviewed scientific papers on food consumer issues in Web of Science has more than tripled during the last decade, together with a substantial increase in the number of journals, research groups and scientists active in the field.

This book with its focus on food consumer science in the Western Balkan countries is a welcome addition to current knowledge base. Besides providing an overview of theoretical and economic foundations for food consumer science, it also illustrates qualitative and quantitative food consumer research applied to sectors and product groups with a high degree of relevance to the Western Balkan countries, a region that stayed previously undeserved out of the picture. Yet, justifications for a stronger focus on food consumer science in this region are numerous. Food and beverages occupy a high share of household expenditure, up to 30–40% in some cases as documented in this book, which underscores the importance of the economic and social dimensions of food choice. The strong gastronomic cultural heritage of the region provides many excellent quality traditional foods, though unknown to most EU consumers and therefore leaving a lot of potential unexploited. Food choice motives and barriers display similarities as well as important dissimilarities as compared to other regions of the EU (Milosevic et al. 2012); measurement scales and constructs may have to be operationalised differently, and other moderators and mediators may emerge in different degrees when explaining food choice. Last but not least, local economic, social, political, technological and structural forces in the food marketing environment shape a specific frame for food production, product differentiation, marketing strategies and public food policy.

The case studies presented in this book cover some highly topical products and industries from the European perspective. While functional foods have been an interesting case for consumer research for almost two decades (Verbeke 2005), this interest gained additional momentum following the approval of the EU regulation 1924/2006 on nutrition and health claims made on foods (Verbeke et al. 2009). Organic and traditional foods in turn are two of the cornerstones of the present EU agricultural quality policy. The establishment of the new EU organic logo and the growing interest for quality indications, such as the Traditional Speciality Guaranteed (TSG) label, as part of the food quality policy illustrate the topicality of this focus. In each of these cases, consumer issues were at the basis of the recent evolutions and shift of emphasis. European consumers have shown a growing interest in food quality rather than quantity. European food policy advocates the overall objective to ensure a high level of protection of European consumers (e.g. from exposure to inferior quality food and misleading information), to provide consumers with the necessary information to make informed (and when possible, also healthier) choices, to guarantee reliable quality food and to foster fair competition and strengthen competitiveness across the European food industry. A better understanding of consumer awareness, attitudes and behaviour, as well as insight in stakeholder and industry positions, and in the role and impact of market structures and macro-environmental conditions is key for successfully reaching such objectives.

European studies in food consumer science consistently show that the classical and readily identifiable socio-demographic characteristics like age, gender, household composition and living environment explain limited variance in food choice. Cross-cultural differences in food choice and its determinants are substantial across Europe. With respect to foods with nutrition and health claims, for example, familiarity with the product, the ingredient and the claim, as well as the perceived personal relevance of food and health have been shown to be important determinants of product acceptance, yet little is known about how health claims and symbols actually influence food choice (EUFIC 2012). Meanwhile, the identification of differences between consumers and countries in their response to foods with health claims has been acknowledged as a challenge for research and policy making. It has also been demonstrated that a European consumer-based definition of traditional foods contains both commonly shared elements, such as an association with a long existence and heritage or a high degree of familiarity and authenticity, as well as elements not equally shared across cultures, for example, the association with special occasions, the notion of locality or naturalness (Vanhonacker et al. 2010). In a similar vein, the marketplace success of organic foods differs to a great extent depending on the country or region. While the role of classical socio-demographic characteristics in explaining cross-cultural variance is rather limited (Aertsens et al. 2009), the potential role of other personal factors, such as attitudes, norms, values, motives and differences, in the macro-environment deserves further attention as explanatory variables for organic food consumption. While consumer insights from a selected number of countries may suffice to understand some of the basic principles at stake, further validation in specific environments and cross-cultural settings is required to complete the picture. This underscores the need for extending food consumer science expertise, insights and networks beyond the borders of countries or regions traditionally involved or covered in previous studies.

The activities performed by the FOCUS BALKANS consortium, part of which are documented in the present book, offer consumer scientists all over Europe and beyond an excellent opportunity to become acquainted with some of the specificities of food consumer science and food choice in Western Balkan countries, as well as with the expertise of the research groups active in this particular science domain in these new EU member states, acceding (or potential) candidate countries. This volume herewith forms a valuable resource for food consumer science scholars seeking to expand their personal expertise and scientific networks.

References

Aertsens J, Verbeke W, Mondelaers K, Van Huylenbroeck G (2009) Personal determinants of organic food consumption: a review. Br Food J 111:1140–1167

EUFIC (2012) Can health claims and symbols lead to healthier eating habits? CLYMBOL press release, http://www.eufic.org/jpage/en/page/PRESS/fftid/clymbol-healthclaims-symbols. Accessed 27 Sept 2012

Milosevic J, Zezelj I, Gorton M, Barjolle D (2012) Understanding the motives for food choice in Westerns Balkan Countries. Appetite 58:205–214

Vanhonacker F, Verbeke W, Guerrero L, Claret A, Contel M, Scalvedi L, Zakowska-Biemans S, Gutkowska K, Sulmot-Rossé C, Raude J, Gransli BS, Hersleth M (2010) How European consumers define the concept of traditional food: evidence from a survey in six countries. Agribusiness 26:453–476

Verbeke W (2005) Consumer acceptance of functional foods: socio-demographic, cognitive and attitudinal determinants. Food Qual Prefer 16:45–57

Verbeke W, Scholderer J, Lähteenmäki L (2009) Consumer appeal of nutrition and health claims in three existing product concepts. Appetite 52:684–692

Acknowledgments

This book is the result of a 3-years European research project funded by the European Commission and the associated countries. The FOCUS BALKANS project was granted from 1 October 2008 until 30 September 2012 (Grant Number: 212 579), with more than 75 researchers collaborating in 14 countries including Bosnia and Herzegovina, Croatia, Former Yugoslav Republic of Macedonia, Montenegro, Serbia and Slovenia. The challenge was to improve existing networks and to establish a new network of scientists, private companies, nongovernmental organisations and public bodies competent in the topic of food consumer science.

Thanks to the effort of the whole team, the research was conducted in an excellent way, respecting the quality and the timeline. A big thank goes to all researchers having made this possible. The preparation of the project was a very hard work and was essentially made possible by Pascal Bernardoni, SEEDEV, Magali Estève, AGRIDEA, and Martine Zaouche-Laniau, GEM. Without their knowledge and networks, this project would not have been born, and we express our gratitude to them especially.

Editing the whole book was an enormous work done by Loredana Sorg, junior researcher at ETH, and we would like to congratulate and thank her warmly here.

We hope this book will support researchers and students, and we wish all a very active learning-by-doing process. This was reached within the FOCUS BALKANS project and has clearly shown that it is the most efficient way to acquire advanced skills in the field of food consumer science.

Contents

Contributors

Corinne Amblard M.Sc. UR CALITYSS, Clermont Université, VetAgro Sup, Clermont-Ferrand, France

Dominique Barjolle, Ph.D. Institute for Environmental Decisions (IED), ETH Zürich, Zürich, Switzerland

Pascal Bernardoni, M.Sc. SEEDEV in Serbia, Belgrade, Serbia

Ružica Butigan, M.B.A. Faculty of Economics & Business, University of Zagreb, Zagreb, Croatia

Miljan Cvetković, Ph.D. Faculty of Agriculture, University of Banja Luka, Banja Luka, Republik Srpska-Bosnia-Herzegovina

Jasna Milošević Đorđević, Ph.D. Ipsos Strategic Marketing, Belgrade, Serbia

Georges Giraud, M.D., Ph.D. UMR CESAER INRA-AgroSup Dijon, DIJON cedex, France

Matthew Gorton, Ph.D. Newcastle University Business School, Newcastle upon Tyne, UK

Cristina Mora, Ph.D. Department of Food Science, University of Parma, Parma, Italy

Boban Mugoša, Ph.D. Institute of Public Health-Montenegro, Podgorica, Montenegro

Mitchell Ness, Ph.D. Newcastle University Business School, Newcastle upon Tyne, UK

Galjina Ognjanov, Ph.D. Faculty of Economics, University of Belgrade, Belgrade, Serbia

Jure Pohar, Ph.D. Department of Animal Science, Biotechnical Faculty, University of Ljubljana, Domžale, Slovenia

Elise Prugnard, M.Sc. UR CALITYSS, Clermont Université, VetAgro Sup, Clermont-Ferrand, France

Natasa Renko, Ph.D. Department of Marketing, Faculty of Economics & Business, University of Zagreb, Zagreb, Croatia

Sanda Renko, Ph.D. Department of Trade, Faculty of Economics & Business, University of Zagreb, Zagreb, Croatia

Burkhard Schaer, Ph.D. ECOZEPT, Freising, Germany

Siet J. Sijtsema, Ph.D. LEI, Part of Wageningen UR, Research Area Consumer & Behaviour, Wageningen, The Netherlands

Igor Spiroski, M.D., M.P.H. Department of Physiology and Monitoring of Nutrition, Institute of Public Health, Skopje, Republic of Macedonia

Žaklina Stojanović, Ph.D. Department of Economic Policy and Development, Faculty of Economics, University of Belgrade, Belgrade, Serbia

Dragana Tar, M.A. SEEDEV in Serbia, Belgrade, Serbia

Ante Vuletić, M.A. Faculty of Economics & Business, University of Zagreb, Zagreb, Croatia

John White, Ph.D. School of Management, Plymouth University, Plymouth, Devon, UK

Martine Zaouche-Laniau Sté GEM (Etudes et Stratégies pour l'Agroalimentaire), Paris, France

Iris Žeželj, Ph.D. Faculty of Philosophy, Belgrade University, Belgrade, Serbia

Karin L. Zimmermann, B.Sc., R.M. LEI, Part of Wageningen UR, Research Area Consumer & Behaviour, The Hague, The Netherlands

Goran Zivkov SEEDEV in Serbia, Belgrade, Serbia

Abbreviations

AHA	American Heart Association
AHC	Agglomerative Hierarchical Clustering
ANOVA	Analysis of Variance
BMI	Body mass index
CA	Cluster analysis
CFA	Confirmatory factor analysis
CFI	Comparative fit index
CNDC	Chronic non-communicable diseases
CVD	Cardiovascular diseases
€	Euro
EFA	Exploratory factor analysis
EFSA	European Food Safety Authority
EU	European Union
EU-27	Today's (2012) 27 member states of the European Union
EuroFIR	European Food Information Resource Network of Excellence
FAO	Food and Agriculture Organization
FCQ	Food Choice Questionnaire
FDI	Foreign Direct Investment
FGD	Focus Group Discussion
FYROM	Former Yugoslav Republic of Macedonia
GDP	Gross Domestic Product
HNI	Household Net Income
HoReCa	Hotel, Restaurant and Catering
IDI	In-depth interview
IPH	Institute of Public Health
N&H claim	Nutrition and Health claim
NGOs	Non-governmental organisations
PBC	Perceived behavioural control
PCFI	Parsimonious comparative fit index

PDO	Protected Designation of Origin
PGI	Protected Geographical Indication
PPP	Purchasing Power Parity
PSUs	Primary sampling units
RMSEA	Root mean square error of approximation
RU	Random utility
SDR	Standardised death rate
TPB	Theory of planned behaviour
TRA	Theory of reasoned action
TSG	Traditional Speciality Guaranteed
UK	United Kingdom
USA	United States of America
WBC	Western Balkan Countries Organization
WHO	World Health Organisation
WTO	World Trade Organization

Chapter 1
Introduction

Dominique Barjolle and Jure Pohar

Abstract Food consumer science is a relatively young science, and research in this field in the Western Balkans is rare. It aims at investigating the complex interactions between the consumers, the marketplace and the industry. These three pillars are explored through several different disciplines and most commonly through the combination of qualitative and quantitative methods.

This book, which addresses scientists and students, explores the main methods, models and approaches of food consumer science applied to six countries of the Western Balkans. It aims at explaining these methods by illustrating each of them with concrete case studies. In the Western Balkans, the proportion of household expenses spent on food is higher than in other European countries, and the changes on the domestic markets present the risk to marginalise many farmers, producers and processors. Therefore, the FOCUS BALKANS project was initiated, and the research conducted between 2008 and 2011 has provided an excellent database for exploring recent changes and trends in food consumption. This opening chapter presents the theoretical economic background for food consumer science, gives a short overview of food consumption in the Balkans and outlines the general conception of the book.

Keywords Food consumer science • Economic theory • Food market • Motives of consumption • Future trends • Niche market

---

D. Barjolle, Ph.D. (✉)
Institute for Environmental Decisions (IED), ETH Zürich,
Sonneggstrasse 33, SOL C2, CH-8092 Zürich, Switzerland
e-mail: barjolle@ethz.ch

J. Pohar, Ph.D.
Department of Animal Science, Biotechnical Faculty, University of Ljubljana,
Groblje 3, 1230 Domžale, Slovenia
e-mail: jure.pohar@bfro.uni-lj.si

D. Barjolle et al. (eds.), *Food Consumer Science: Theories, Methods and Application to the Western Balkans*, DOI 10.1007/978-94-007-5946-6_1,

1.1 Preliminary Considerations

Food consumer science is relatively young. It could be considered as a synonym for or a hybrid of two distinct sciences. On the one hand, there is a part that might be regarded as 'hardware', that is, science about food or *food science*, while on the other hand there is a part that might be regarded as 'software', namely science about consumers or *consumer science*. The latter emerged in the 1960s with a growing consideration by firms interested in improving the marketing management process (Engel et al. 1995).

Food consumer science thus intends to combine these two sciences and pursues a holistic approach towards hardware (referring in particular to natural sciences such as chemistry, biochemistry, microbiology, process technique, etc.) and software (i.e. social and humanistic sciences, mostly sociology and psychology). The latter should tell us why, when and how the consumer will consume food, while the first should examine how food is produced and processed. In addition, it could include a study of the impacts of food on the growth, development and health of the human being. If food science was restricted to the mere study of the production of food, it would be more correct to speak about food production science which – when studying the food industry in particular – could also be called food technology science. When food is not studied as a consumer/consumable product but rather as a factor influencing growth, development and health, and where the consumer is considered merely one of the living organisms (higher mammals), ignoring the psychological and focusing only on the physiological component, both parts – that is, food science and consumer science – meet in an interface usually designated as *nutrition science*.

Food consumer science aims at investigating the complex interactions between consumers and industry in the framework of the food marketplace. This complex system is explored through different disciplines like economics, food science, nutrition, psychology, sociology, marketing and anthropology (Steenkamp 1997): the research is based on several models and theories (Table 1.1). The scope of consumer science goes beyond understanding the willingness of consumers to buy. Consumer science is indeed a social science, which aims at increasing general social welfare by taking into account the deep aspirations of the consumers for developing more convenient products and marketing tools (Solomon et al. 2006). Of course, the consumer is the main subject of research, but the interactions with the two other dimensions (market and industry) are fundamental to be able to understand and forecast consumption patterns and consumers' behaviour. Consumer research is not only the testing of new products or the evaluation of the prices the consumers can pay for a product. It is also related to forecasting and investigating prospects in research and development activities for companies and therefore closely linked to the elaboration of products and marketing innovations. Qualitative and quantitative methods are used, most often in combination, like in other social sciences.

This book explores the main methods, models and approaches used in food consumer science with applications within the Western Balkans. Three aspects render this publication unique: first, the systematic presentation of the most useful methods

Table 1.1 Main disciplines, theories and models for food consumer science (not exhaustive list)

	Main discipline	Theory or model
Food consumer science	Economics	Asymmetry of information
		Economy of quality
		Food safety economics
		Institutional economics
		Economic household models
	Food science/nutrition	Biopsychological approach
		Health belief model
	Psychology	Theory of reasoned action
		Theory of planned behaviour
	Sociology	Food choice
		Process model
	Communication	Theory of linear knowledge transfer or demand driven
	Marketing	Food supply chain management
		Corporate social responsibility

to explore food consumption; second, a focus on research results related to consumers within the Western Balkans; and third, a consideration of the four relevant markets for developing healthy food consumption (health claim, organic and traditional products and fruits).

This book presents the results of the European research programme FOCUS BALKANS, which was conducted between 2008 and 2011 by a large consortium of 17 partners (see Chap. 1.6).

The quality of the research findings was continuously checked by a steering committee of eight scientists, who are co-authors of this book. External supervision was carried out by Agata Pienadz, scientific officer at the Directorate for Research and Innovation at the European Commission. Two external reviewers, Jutta Roosen, professor at the Technische Universität München in Germany, and Monique Raats, University of Surrey in the United Kingdom (UK), undertook the midterm review of the project. Thanks to their collaboration and the fruitful comments of three external reviewers this book was possible, and we would like to acknowledge the commitment of all these experts.

The rest of this chapter introduces the book and provides, in turn, the theoretical anchorage of food consumer science from an economic perspective, a short overview of food consumption in the Western Balkans and the outline of the structure of the whole book.

1.2 Economic Foundations for Food Consumer Science

Classical and neoclassical microeconomic perspectives focus on the functioning of markets and are based on several assumptions related to demand and, inductively, to 'consumer choice'. These theories consider the optimal allocation of resources under conditions of perfect information. A main assumption is that the consumer has a free choice in terms of whether to allocate her/his resources to one product or alternatively to another one. Further, it is assumed that markets converge to equilibrium usually through a 'trial-and-error' process.

Recent economic theory highlights the importance of considering more carefully the consumer as a research object and how markets may fail due to information asymmetry. Akerlof (1970) studied the link between quality and uncertainty. His approach to uncertainty is based on the assumption that, in a certain context of trade such as second-hand cars, uncertainty is inherent to the nature of the transaction itself. Asymmetry of information about products is essentially linked to *experience goods* (Nelson 1970), for which the consumer assesses the quality only when she/he actually consumes the product, and *trust goods* (Darby and Karni 1973), sometimes known as *credence goods*, for which the quality cannot be assessed directly by the consumer either prior to or post purchase (e.g. for food: production methods or nutrition content). Shapiro (1983) demonstrated even that 'premium prices' are not in contradiction with the general *welfare theory* because they are necessary to cover reputation costs (certification and control costs can also be added to it), in which any firm has to invest in order to gain the trust of consumers.

The importance of food safety and related economic concerns should be systematically considered in food consumer research. Recent analyses showed that the values of safety, nutrition, taste and price were amongst the most important for consumers (Lusk and Briggeman 2009). As showed by Caswell (1998), food labelling reduces information asymmetry. Consequently, the economics of food safety is included into the theoretical framework of contemporary food consumer science. Food safety issues cover how farmers produce food (what chemicals they use when growing plants and how they feed their animals), how food is processed, how it is sold and what sort of information is provided on product labels.

Economists often suggest an institutional perspective for research on consumer behaviour. Food consumer science has to refer to *institutional economics*, and the roles of different players in the market should be analysed in the context of understanding contemporary food consumer behaviour. This rich institutional context includes consumers' associations, farmers and processors, retailers and other sellers, regulatory institutions and governmental bodies in the area of public health, food regulation, trade, innovations and technology, education and information. Derived from an institutional economics perspective, the *economics of quality* (Karpik 2007) studies the social process of the setting up of quality norms, that is, the negotiation and agreements' reaching process. This perspective highlights the balance between offers and demands beyond individual instantaneous transactions and through the broader understanding of the social and institutional context, which is evolving at any point in time. This theory is a must to explore the development of the quality of food along the supply chain.

Finally, the *economics of households* provides useful insights particularly in the context of transition and developing countries. This approach is very complete and useful in many situations and contexts and is detailed as a systemic step-by-step approach at the local, national or regional (in the sense of a group of countries) levels to describe the markets and the food habits and to define the priorities of any food policy. Nevertheless, the objectives of the FOCUS BALKANS project were not related to that topic, and this approach is not described further in this book.

Starting from these perspectives, food consumer science considers the consumer as a complex object to be investigated from different perspectives. Increasing health costs have led to a growing interest from public health agents, and the second next user of the results are therefore public health specialists and policymakers. The growing emphasis on establishing integrated public policies between agriculture, food and health and for involving the private sector to fulfil corporate social responsibilities is opening a very promising future for the young field of food consumer science.

1.3 Food Consumption in the Western Balkans

During the socialist era (from 1945 to 1991), the Western Balkan countries (WBC) covered by this research were part of Yugoslavia, which engaged in a policy of non-alignment to the EU and USSR. The disintegration of the former Yugoslav states continues to colour the relationships between successor states.

Consumer science is of major importance to improve the functioning of food markets. These countries are facing the liberalisation of the domestic markets and experiencing greater global competition. Understanding consumers' behaviour and expectations is pivotal for their future development.

In the context of accession to the European Union (EU), all these countries have undertaken reforms to harmonise their legislation with EU directives and regulations. The World Trade Organization (WTO) accession process also triggered reforms. Slovenia joined the EU in 2004, and Croatia will join on 1 July 2013. FYROM, Montenegro and Serbia are candidates for membership of the EU, while progress in Bosnia and Herzegovina has been slower, although the country hopes to gain candidate status by 2014.

The harmonisation with the EU framework implies a greater involvement of non-governmental stakeholders and supranational actors in food and nutritional policy.

Food markets in WBC are strongly influenced by the legacy of the former Yugoslavian agro-industrial sector, in which the *agro-kombinats* (socially owned complex gathering primary production and processing units) mostly supplied the formal marketing channels. The concept of social ownership prevailed in Yugoslavia and corresponds to the one of inalienable collective ownership. With the privatisation of the *kombinats*, the emergence of new types of producers and the process of EU accession, legislations and regulations have focused more and more on product quality to the benefit of consumers. The general pattern of food consumption in the WBCs can be compared against those for existing EU member states. The total expenditure on food in the WBC is on average twice that of the average for the EU-27 (Table 1.2). Yet the proportion of household expenses accounted for by food is significantly higher in the WBC, although this is less marked in Slovenia and Croatia than in the other countries of the region, which are not yet members of the EU. The percentage of household expenses allocated to transport, communication,

Table 1.2 Average household expenditures in euros, 2005

2005	Food and non-alcoholic beverages	Alcoholic beverages, tobacco and narcotics	Clothing and footwear	Housing, water, electricity, gas and other fuels	Furnishings, household equipment and routine maintenance of the house	Health	Transport	Communications	Recreation and culture	Education	Restaurants and hotels	Miscellaneous goods and services	Total
European Union (27 countries)	3,594	560	1,412	6,947	1,416	796	3,078	738	2,187	238	1,417	2,291	24,674
European Union (25 countries)	3,675	577	1,483	7,310	1,493	832	3,252	768	2,313	250	1,500	2,426	25,879
European Union (15 countries)	3,851	626	1,648	8,127	1,667	907	3,649	807	2,594	273	1,716	2,744	28,609
Slovenia	3,966	575	1,678	5,483	1,389	356	3,717	950	2,234	202	1,035	2,220	23,805
Croatia	4,564	548	1,059	4,983	697	315	1,484	729	853	105	465	1,039	16,841
FYROM	5,545	453	994	1,496	622	414	870	590	479	103	543	511	12,620
In percentage													
European Union (27 countries)	14.6	2.3	5.7	28.2	5.7	3.2	12.5	3.0	8.9	1.0	5.7	9.3	100.0
European Union (25 countries)	14.2	2.2	5.7	28.2	5.8	3.2	12.6	3.0	8.9	1.0	5.8	9.4	100.0
European Union (15 countries)	13.5	2.2	5.8	28.4	5.8	3.2	12.8	2.8	9.1	1.0	6.0	9.6	100.0
Slovenia	16.7	2.4	7.0	23.0	5.8	1.5	15.6	4.0	9.4	0.8	4.3	9.3	100.0
Croatia	27.1	2.4	7.0	23.0	5.8	1.5	15.6	4.0	9.4	0.8	4.3	9.3	100.0
FYROM	43.9	3.3	6.3	29.6	4.1	1.9	8.8	4.3	5.1	0.6	2.8	6.2	100.0

Source: Eurostat (http://epp.eurostat.ec.europa.eu)

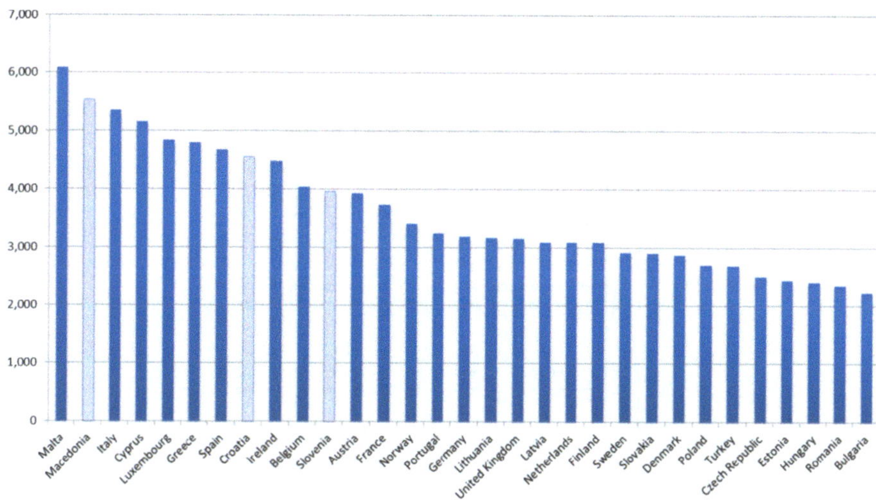

Fig. 1.1 Average household expenditures for food and non-alcoholic beverages in euros, 2005. Data of WBC available only for FYROM, Croatia and Slovenia (*Source*: http://epp.eurostat.ec.europa.eu)

recreation/culture, education, restaurants/hotels and miscellaneous items is much lower in the WBC than the EU average.

The total expenditures in euros for food and non-alcoholic beverages in Croatia, FYROM and Slovenia are above the average of the EU (Fig. 1.1). This shows in part the importance of food for the local economy and in the local way of life.

A significant growth in household income and expenditure was recorded in most of the WBC until the economic crisis beginning in 2008. At the same time, changes have occurred both in the structure of income and expenditures. Both prior to and after the recent financial crisis, Slovenian households have been relatively better off, and the proportion of household expenditure accounted for by food is markedly less (17.6%) than for the other countries studied in the project. In all other countries studied, this percentage is significantly higher (32–43% range). The richer a country is, the higher is the living standard of its population and the lower the percentage of household income spent on food. The highest proportion spent on food takes place in FYRO Macedonia, then in Serbia and the smallest in Slovenia (Table 1.3).

In order to maintain and increase market participation, the Western Balkan food industry (farmers, processors, producers and retailers) must comply with increasingly demanding global requirements, for example, sanitary regulations. Nevertheless, these requirements and evolving market conditions may force many small economic agents (farmers, processors and producers) burdened with unfavourable sizes or structures of farms, the fragmentation of holdings, and the low level of specialisation to close their businesses, unless they find the means to invest in modern technologies

Table 1.3 Average household expenditures (percentage)

	Slovenia		Croatia		Bosnia and Herzegovina	Serbia		Montenegro		Macedonia	
In percentage	2007	2010	2007	2010	2007	2007	2010	2007	2010	2007	2010
Food and non-alcoholic beverages	18	17.6	32.6	32.1	31.9	38.3	41.3	35.4	39.8	41.7	43.1
Alcoholic drinks and tobacco	2.8	2.4	3.7	3.6	3.3	4.7	4.4	3.5	3.5	4.3	4.1
Clothes and footwear	7.7	7.6	8	7.3	5.1	6.7	4.8	8.2	7.4	7.6	6.5
Housing, water, electricity and other fuels	12.8	13.9	13.7	14.4	22.1	15.2	16	12.9	15.4	10.4	12.8
Furniture, household equipment and maintenance	7.4	7.7	5.1	5.2	5.4	5	4.4	5.1	3.9	6.2	5.4
Healthcare	1.8	2.4	2.5	2.8	4	4.3	4.1	4	3.3	3	3.8
Transport	18.1	16.5	11.4	11	11.2	11.2	9	11	8.7	8.4	6.5
Communications	5	5.1	5.1	5.1	3.3	3.3	3.9	6.1	5.1	4.5	4.1
Recreation and culture	10.3	10.6	6.2	6	3.6	4.4	4.7	3.9	3.7	4.1	2.7
Education	1.1	0.9	0.7	1	0.5	0.8	0.9	2.3	1.8	0.9	1
Restaurants and hotels	4	4	3.3	3.1	2.8	1.7	1.9	2.7	2.3	4.9	5.1
Other goods and services	11	11.3	7.7	8.4	6.4	4.4	4.6	5.1	5	4	4.9
Personal consumption – total	100%										

Source: Milosevic et al. 2009

and adapt to the new environment. The changing nature of domestic markets presents the risk of marginalising many farmers, producers and processors due to the increasing market share of international and major domestic supermarkets, the introduction of standards, new business practices and consumers' increased demand for quality and safe foods. Throughout the WBC, small-scale agricultural producers particularly in marginal areas are often old and with basic education.

In this context, the FOCUS BALKANS project was initiated and the research conducted between 2008 and 2011. It provides an excellent database for exploring in-depth recent changes and trends of food consumption in six WBC (Bosnia-Herzegovina, Croatia, FYRO Macedonia, Montenegro, Serbia and Slovenia).

1.4 The FOCUS BALKANS Project

The spirit of the FOCUS BALKANS project was to interlink research, training and networking activities regarding food consumer science so as to boost research in a sustainable way: all six countries conducted individual research studies on niche markets and a quantitative consumer survey, participated in and/or organised trainings (6), networking meetings (32) and open seminars (2). Even after the end of the project, the participant researchers' network continues to develop activities in that field.

The general objective of FOCUS BALKANS was to improve competencies and understanding in the field of food consumer science in the WBC and especially to:

- Have a better knowledge and understanding of WBC food consumers, with a focus on products with positive nutritional properties (food with 'health claim' and fruits) and sustainability claims (organic and traditional food).
- Develop competencies and understanding of public organisations, private enterprises and NGOs regarding food consumer science in the WBC (educational trainings).
- Develop a network of universities, institutes, high schools, consumer organisations, NGOs and private enterprises active in the field of consumer science related to food (networking meetings and open seminars).

The most important findings were that in general more promotion for food in relation to health promotion (improvement of the diet) must be undertaken by governments and that supply chains (from producers and processors to small retailers) must be fostered and better structured. Voluntary food standards must be better communicated to consumers and more clearly tagged and positioned. The food industry, policymakers and researchers desire to be active within scientific and professional networks in order to continue conducting actively such research.

The project activities resulted in the significant participation of WBC scientists in projects related to food consumer science. More than 100 scientists from the region were directly involved in research, training, networking or dissemination activities. Furthermore, they gained competencies in presenting scientific papers at conferences and in publishing papers in scientific journals in order to make food consumer science in the WBC more broadly known amongst the scientific community (see on www.focus-balkans.org).

The WBC partners have elaborated new methods, which fit the specific WBC context. They are now able to apply them to further scientific studies and concrete programmes. Knowledge has been disseminated through several bachelor and master programmes for which this book constitutes an excellent teaching resource.

1.5 Objectives, Scope and Outline of the Book

This book presents major methods used in food consumer science, illustrating each of them with concrete results obtained through the European research programme FOCUS BALKANS. The scope of this book is essentially market study, motives of consumption, future trends and challenges for niche food markets and consumer preferences. The scope was defined according to the most preferred and needed research results by the private and public sectors in the Balkans. This was assessed at the beginning of the project by the steering committee members based on their experience and knowledge of the state of the art for research in the region.

This book is addressed especially to scientists and students. The data and main conclusions are of particular interest for policymakers concerned with regulating the food supply chain, public authorities in charge of controlling agriculture, food production and the food market, civil society organisations charged with defending consumers' rights and for private enterprises within the agricultural and food sector.

The seven core chapters on methods and results provide a thorough understanding of both the main theories that inform consumer food science and the methods that can be used to gain an empirical understanding. Since it is not possible to explore all food market segments in the scope of this book, the different chapters illustrate methods with examples from products with health and nutritional claims (Chap. 5), the market for fruits (Chap. 6), organic products (Chap. 7) and traditional foods (Chap. 8). As quantitative and qualitative approaches are used in food consumer science, both approaches are presented (qualitative methods in Chaps. 5 and 6; quantitative methods in Chaps. 3, 4, 7 and 8).

The application of techniques draws on a major piece of cross-national research. Quantitative methods are illustrated from results of a quantitative survey conducted in all six WBC included within the project. The cross-national survey covered 3,085 adult respondents. The respondents completed the Food Choice Questionnaire (FCQ) (Steptoe et al. 1995) as well as additional questions on food choice in each mentioned country, and the sample was stratified to match key characteristics of the population. The survey sought to understand the motives for food choice in the WBC and to segment and profile groups of consumers. This approach formed the basis of consumer profiling and is presented in Chap. 3. Chapter 4 presents the method of cluster analysis based on the results of the FCQ.

Table 1.4 Scope, methods and illustrations presented in the book

Scope	Method	Illustration
Market study	In-depth interviews with experts	Health claim products
Motives for consumption	Consumer focus groups	Fruits
Future trends and challenges in niche food markets	Delphi method	Organic
Consumer preferences	Conjoint analysis	Traditional products

Methods and results about each topic of the research (fruits, health claim products, organic and traditional food) have been described according to the matches between scope/methods and illustrations described in Table 1.4.

At the end of each chapter, a discussion of the method presents its benefits and utility as well as the main difficulties that may be encountered in its application. Altogether, the chapters cover the major issues needed for a scholar to become familiar with food consumer science, both in terms of methods and markets.

1.6 Acknowledgement

The content of this book is based on the findings of the FOCUS BALKANS project. We would like to thank all the researchers and institutions who contributed to the FOCUS BALKANS project and thus made this publication possible.

The researchers and institutions involved were

Institution	Researcher
ETH Zürich (coordinator of the project)	Dominique Barjolle
	Delphine Eynaudi
AGRIDEA	Magali Estève
	Sophie Réviron
	Astrid Gerz
SEEDEV	Pascal Bernardoni
	Josip Jagust
	Dragana Tar
	Goran Zivkov
Ipsos Strategic Marketing – SMMRI	Hana Baronijan
	Jasna Milošević Đorđević
	Dragisa Bjeloglav
	Ana Delic
	Sonja Stancic
	Tanja Stojadinovic
	Dalibor Misic
	Marija Tadic
	Jelena Matovic
	Jelena Jakic
	Srdjan Bogosavljevic

(continued)

(continued)

Institution	Researcher
DLO LEI Wageningen UR	Abele Kuipers
	Siet Sijtsema
	Karin Zimmermann
	Mariet van Haaster-de Winter
	Isabelle van den Berg
	Niels Tomson
	Barbara van der Hout
	Harriette Snoek
	Xiaoyong Zhang
VetAgroSup	Corinne Amblard
	Georges Giraud
	Julie Mardon
	Elise Prugnard
University of Newcastle, United Kingdom	Matthew Gorton
	John White
	Mitchell Ness
ECOZEPT	Burkhard Schaer
	Nina Berner
GEM	Martine Zaouche-Laniau
	Michel Timsit
University of Parma, Italy	Cristina Mora
	Maria Cecilia Mancini
	Davide Menozzi
	Giovanni Sogari
	Dario Caccamisi
	Carla Campana
University of Ljubljana, Biotechnical Faculty, Slovenia	Marija Klopcic
	Jure Pohar
University of Zagreb, Faculty of Economics, Croatia	Natasa Renko
	Sanda Renko
	Ruzica Butigan
	Ante Vuletic
Faculty of Economics, University of Belgrade, Serbia	Žaklina Stojanović
	Galjina Ognjanov
	Radmila Dragutinović-Mitrović
	Aleksandra Zečević
	Dragan Stojković
	Jelena Filipović
	Svetlana Popović
	Mirjana Gligorić
Faculty of Philosophy, University of Belgrade, Serbia	Iris Žeželj
	Nebojša Petrović

(continued)

(continued)

Institution	Researcher
University of Banja Luka, Faculty of Agriculture, Bosnia	Miljan Cvetkovic
	Lidija Tomić
	Aleksandar Figurek
	Dimitrije Marković
Institute of Public Health, Montenegro	Boban Mugosa
	Ljiljana Zizic
	Borko Bajic
	Sanja Scepanovic
	Zorica Đorđević
	Balša Radulović
Institute for Health Protection, Macedonia	Vladimir Kendrovski
	Igor Spiroski
University of Wageningen, the Netherlands	Ivo van der Lans
	Cees De Graaf

References

Akerlof F (1970) The market for lemons: quality uncertainty and the market mechanism. Q J Econ 84:488–500

Caswell J (1998) How labeling of safety and process attributes affects markets for food. Agric Resour Econ Rev 27(2):151–158

Darby M, Karni E (1973) Free competition and the optimal amount of fraud. J Law Econ 16:67–88

Engel JF, Blackwell RD, Miniard PW (1995) Consumer behaviour, 8th ed. The Dryden Press, Fort Forth

Karpik L (2007) L'économie des singularités. Ed. Gallimard, Paris

Lusk JL, Briggeman BC (2009) Food values. Am J Agric Econ 91(1):184–196

Milosevic J, Barjolle D, Bernardoni P (eds) (2009) Analysis of the current situation of the Food Consumer Sciences Knowledge System in the Western Balkan Countries. FOCUS-BALKANS, DELIVERABLE N°2.1

Nelson P (1970) Information and consumer behavior. J Polit Econ 78:311–329

Shapiro C (1983) Premiums for high quality products as returns to reputations. Q J Econ 98(4): 659–679

Solomon M, Bamossy G, Askegaard S, Hogg MK (2006) Consumer behaviour: a European perspective. 3rd ed, Prentice Hall, Europe

Steenkamp J-BEM (1997) Dynamics in consumer behaviour with respect to agricultural and food products. In: Wieringa B, Tilburg A van, Grunert K, Steenkamp J-BEM, Wedel M (eds), Agricultural marketing and consumer behaviour in a changing world. Kluwer Academic Publishers, Dordrecht, pp 143–188

Steptoe A, Pollard T, Wardle J (1995) Development of a measure of the motives underlying the selection of food: the food choice questionnaire. Appetite 25:267–284

Chapter 2
Theories of Food Choice

Matthew Gorton and Dominique Barjolle

Abstract This chapter introduces and evaluates the main models that seek to understand consumer behaviour relating to food. Three of the most influential approaches applied specifically to understand food choice are described in detail, and their strengths and weaknesses are discussed: (a) economic (household and random utility) models, (b) the food choice process model and (c) the theory of reasoned action/theory of planned behaviour. The selection of these three groups of models reflects their influence within the food choice literature and use within policy studies. Given the large number of empirical studies that have validated the theory of planned behaviour for the study of food choice, this model was selected as a core theoretical framework for the Focus Balkans project. From reviewing the models, conclusions are drawn relating to the current state of the literature on food choice.

Keywords Food choice • Economic household model • Random utility model • Food choice process model • Theory of planned behaviour

M. Gorton, Ph.D. (✉)
Marketing Subject Group, Newcastle University Business School,
5 Barrack Road, Newcastle upon Tyne NE1 4SE, UK
e-mail: matthew.gorton@newcastle.ac.uk

D. Barjolle, Ph.D.
Institute for Environmental Decisions (IED), ETH Zürich,
Sonneggstrasse 33, SOL C2, CH-8092 Zürich, Switzerland
e-mail: barjolle@ethz.ch

D. Barjolle et al. (eds.), *Food Consumer Science: Theories, Methods and Application to the Western Balkans*, DOI 10.1007/978-94-007-5946-6_2,

2.1 Introduction

This chapter introduces and evaluates the main models that seek to understand consumer behaviour relating to food. As a starting point, several authors seek to list the factors that influence food choice. These factors can be divided into three categories: (1) product, (2) person and (3) environment (or situational context). All these factors can, for example, be found in one of the first models of food choice, presented by Randall and Sanjur (1981). In the latter, these three sets of factors are listed as determinants of food preferences (see Fig. 2.1). Each group of factors includes several characteristics.

It should be noted, however, that Randall and Sanjur's (1981) model lists variables rather than identifies causal relationships and remains largely descriptive in nature. Other, more complex models attempt to provide the basis for explaining, predicting and also controlling food choice by demonstrating causal relationships. These attempts are not confined to one branch of the social sciences. Rather economists, marketers, psychologists, social anthropologists and sociologists have all made significant contributions. Marketers have typically drawn on *generic* models of buyer behaviour to explain food choice such as consumer decision process models and the Howard-Sheth model of buyer behaviour (Sheth 2011). In this chapter, we evaluate, however, three of the most influential approaches applied *specifically to understand food choice*: (a) economic (household and random utility) models, (b) the food choice process model and (c) the theory of reasoned action/theory of

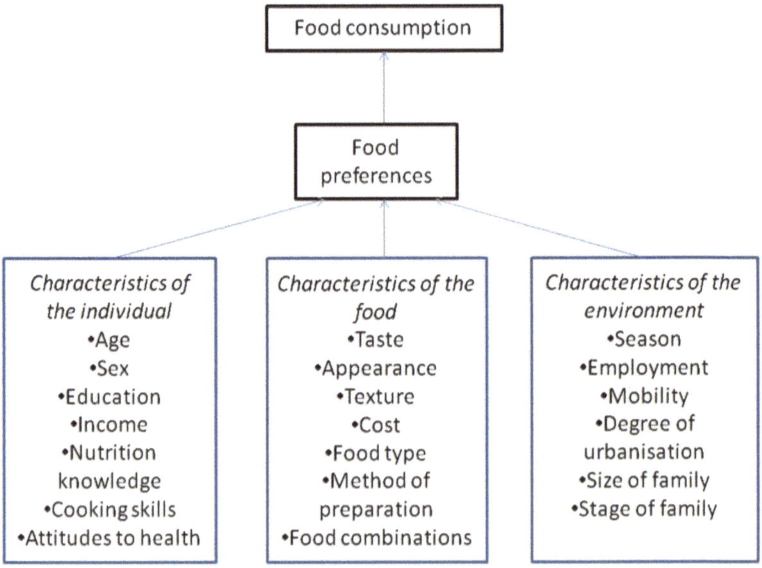

Fig. 2.1 Factors influencing food preferences (*Source*: Randall and Sanjur 1981, p. 154)

planned behaviour. The selection of these three groups of models reflects their influence on the food choice literature and use within policy studies (Grunert et al. 1996; Conner and Armitage 2008). After describing the main features of each approach, we evaluate their strengths and weaknesses in turn.

2.2 Economic Models

2.2.1 Economic Household Model

This approach dates back to the work of Becker (1965), who developed a work–leisure model, where the household is the unit of analysis. He reasoned that households produce utility-yielding commodities (outputs) with combinations of market goods and time. The household can allocate its time to one of three uses: (a) labour market time (generates income to acquire market goods), (b) household production time (e.g. cooking) and (c) consumption time (e.g. eating). In relation to food, meals typically require labour to be allocated to all three uses: labour market time to generate the income to pay for ingredients, production time to cook the meal and consumption time for eating. The uses of time are however competitive with each other – an additional hour spent at work is one less hour that can be devoted to household production or consumption.

Bonke (1992) utilises Becker's (1965) model to analyse the effects of increases/decreases of income and disposable time on patterns of food consumption. The former argues that Western societies witnessed in the post-war period rising incomes and an increase in the number of women in the labour market. The rise in women's working hours led to a decrease in the time available to spend on household activities. Bonke (1992) argues that households with more money but less time to spend on household activities will substitute non-convenience foods for convenience foods, reducing the time spent on cooking.

To test the relationships between disposable time, disposable income and food choice, Bonke (1992) draws on cross-sectional data from Denmark. Foods were classified into non-convenience goods (i.e. ingredients used for cooking from scratch), semi-convenience goods (e.g. ready-made sauces) and convenience goods (i.e. ready meals, meals eaten out of the home in bars and restaurants). Relative expenditure on the three food categories was assessed in terms of the household's disposable and non-disposable income. Households that are both rich and busy were supposed to demand a higher proportion of convenience foods, while those households that are poor with a lot of disposable time would utilise a higher proportion of cheaper, non-convenience foods.

Figure 2.2 indicates that, in accordance with Bonke's (1992) hypothesis, busy and rich households allocate a greater proportion of expenditure on convenience food and a smaller proportion on non-convenience foods. The economic household model has been utilised to explain the rise in the number of restaurants and consumption of ready meals during periods of economic growth and the sensitivity in

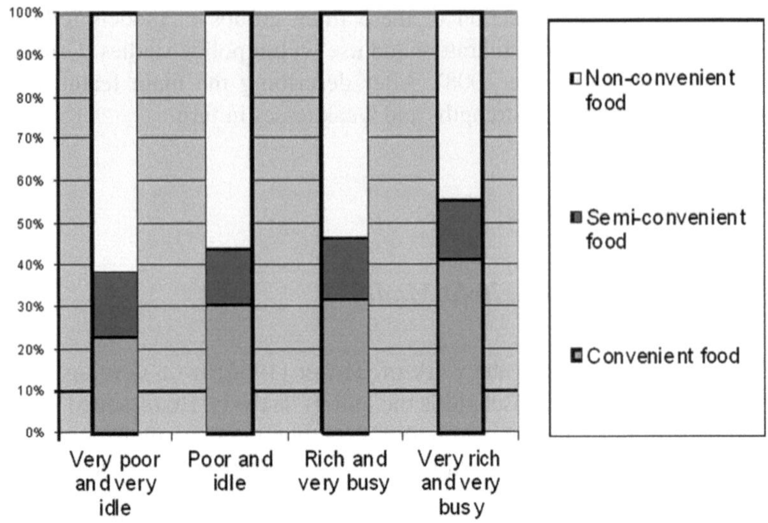

Fig. 2.2 Expenditure on food with different degrees of convenience according to household resources (*Source*: Bonke 1992, p. 48)

the fortunes of restaurants to changes in income (Schumacher and Boland 2005). In keeping with Bonke's approach, the current global recession has been associated, in many countries, with a sharp fall in out-of-home expenditure on food and rising interest in 'grow your own' allotments and vegetable plots (Green 2008). In the Western Balkan Countries (WBC), semi-subsistence farming remains relatively important and plays a significant role in food consumption, especially in rural areas (Lampietti et al. 2009).

The economic household approach, however, presents several limitations. For instance while the cross-sectional data from Denmark fit well with the economic household model, Bonke's approach is less able to explain cross-national variations in the consumption of convenience foods. For example, average incomes and labour market participation are higher in Sweden than Spain, but the share of food spending accounted for by out-of-the-home meals is significantly higher in the latter country (Swoboda and Morschett 2001). Ready-to-cook products for consumption at home (while generally witnessing increasing sales in Western Europe) have greater penetration in the United States of America (USA) and the United Kingdom (UK) compared to France and Italy (Swoboda and Morschett 2001). This pattern cannot be reduced solely to differences in incomes and disposable time. Culture also plays a critical role in explaining cross-national variations in food choice. For instance, to take an extreme example, cross-national variations in the consumption of horsemeat owe little to variations in incomes and disposable time but rather whether its consumption is culturally sanctioned. Sociologists and anthropologists seek to incorporate a wider set of factors into their explanations of food choice.

2.2.2 *Random Utility Models*

Random utility (RU) models consider the choices made by consumers between mutually exclusive discrete alternatives (such as selecting between brands) (Hanemann 1984; Baltas and Doyle 2001). Consumers are considered to maximise utility which is revealed through observed choices (i.e. consumers choose the product with the most desired set of attributes from a set of alternatives). It is assumed that a consumer's utility function is deterministic to that person and thus contains some components which are unobservable to the researcher, for example, taste preferences (Hanemann 1984). The utility function can thus be decomposed into two parts: (a) a deterministic component, specified as a function of measured attributes of alternative products and/or individual characteristics, and (b) a stochastic component, representing unobserved attributes affecting choice, interpersonal variations in utilities arising from heterogeneity in tastes and measurement errors (Baltas and Doyle 2001).

In relation to food choice, RU models have been employed, for example, to analyse choice of organic versus conventional foods (Gracia and de Magistris 2008) and preferences for extra-virgin olive oil (Cicia et al. 2002). In applying RU models, researchers have, however, encountered a number of practical difficulties. First, there tends to be a trade-off between simplicity and realism in the choice alternatives considered (Baltas and Doyle 2001). If the number of alternatives becomes too large, it may not be manageable. A simple design, however, may lead to poor validity. For example, if in studying the choice between different brands of instant coffee, the researcher failed to recognise that each brand is typically available in various pack sizes and caffeinated and decaffeinated versions, estimations are likely to be of limited value. Second, most empirical studies have focused on the mutual exclusivity case, where consumers choose *one* brand or product from a particular category. However, consumers may select a subset of brands from the product category; for example, a consumer may purchase two types of cheese at the same time (Hanemann 1984).

2.3 Food Choice Process Model

Furst et al. (1996) developed the food choice process model. It is one of the most influential approaches to be based on grounded theory and derived from qualitative research. There are three main components of the model: the life course, influences and personal systems (Fig. 2.3).

The life course refers to past and current eating experiences. Furst et al. (1996) argue that to understand current patterns of food consumption, it is necessary to understand trajectories, defined as a person's persistent thoughts, feelings, strategies and action over the lifespan. Devine et al. (1998) contend that trajectories develop in specific situational and historic contexts that become persistent and

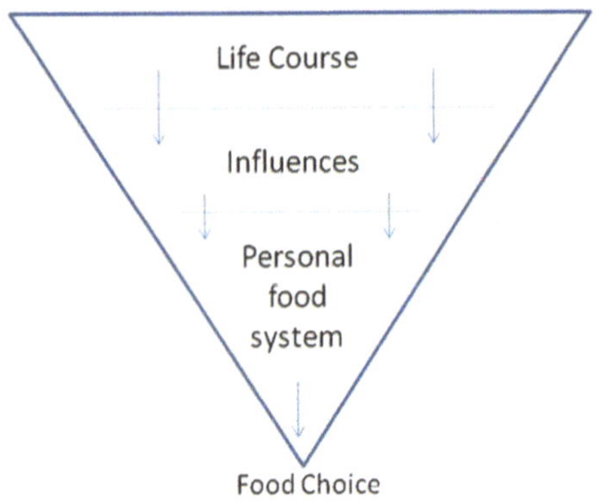

Fig. 2.3 Food choice process model (*Source*: adapted from Sobal et al. 2006, p. 3)

exhibit their own momentum and continuity. The family unit is seen as the most important situational and historic context, so that a person's upbringing moulds patterns of food consumption long after they have left the parental home. However, the authors are careful to avoid crude historical determinism – recognising that researchers should also capture transitions – shifts in a person's life that 'lead to changes or solidify the continuation of behaviours, including food choice patterns' (Sobal et al. 2006, p. 4). For instance, illness may act as a critical transition, disturbing habits (Falk et al. 2000).

In Furst et al.'s (1996) model, influences refer to ideals, resources, social framework and the food context. Ideals are the symbolic meanings people associate with food, such as social status and whether a particular good is regarded as 'proper food'. The authors note that some people are more 'food centred', deriving pleasure, safety and symbolic value from cooking, while others display low 'food salience':

> I don't make an issue out of having a sit down meal or whatever.... I don't think that you have to make like an all out ... effort to make dinner every day. It doesn't have to be like a main function of your life. (Furst et al. 1996, p. 254)

Resources are classified as tangible (money, equipment and space) or intangible (culinary knowledge, skills and time). Both sets are regarded as important determining factors. The social framework captures the nature of interpersonal relationships, social roles and meaning. Families are regarded by Furst et al. (1996) as the most important set of interpersonal relationships influencing food choice, with individuals 'enacting or being assigned particular household food roles' (Furst et al. 1996, p. 255). These roles may conflict with individual preferences, as illustrated in the following quotation, where one interviewee indicated that she placed family needs above her own:

> if it wasn't for them [the family] I probably wouldn't [cook], probably just have apples or something.... I'd probably just be eating one thing ... I love doing it [preparing food], but yeah, for them it's ... fun to do and I do it for them, but it's not a priority for me. (Furst et al. 1996, pp. 255–256)

The third component of the model, the personal food system, relates to mental processes whereby people translate influences on their food choices into how and what they eat in a specific context. This consists of two main components: value negotiations, which involve evaluating the varying merits of different factors, and strategies. Furst et al. (1996) identify six main values pertinent to food choice: sensory perceptions, monetary considerations, convenience, health/nutrition, managing relationships and quality. Strategies capture well-established habits or rules; for example, one woman's rule for buying yoghurt was

> ... there's [a] certain brand of yogurt that my daughter likes ... I will automatically buy that brand. Because I know if I buy the other brand [cheaper] it will just sit in the refrigerator and rot. (Furst et al. 1996, p. 60)

Devine et al. (1998) utilise Furst et al.'s (1996) model to explain patterns of fruit and vegetable consumption. It has also informed studies of the food choice of older consumers (Falk et al. 1996) and newly married couples (Bove et al. 2003).

The food choice process model incorporates a far wider set of factors to explain food choice than the economic household model, and it is designed to be comprehensive (Sobal et al. 2006). For instance, the prominence given to life histories is in contrast to the economic models of Becker (1965) and Bonke (1992), where past behaviour and personal health (illness) do not enter as explanatory variables. However, the model is based on a narrow data set – 29 interviews with adults drawn from New York State (USA). Each interview only lasted for 20–30 min, which is short compared against the norms in qualitative research for in-depth interviews and appears inadequate to capture an individual's life course, influences and personal food systems. While the model seeks to be universal, its validity in a cross-national setting has not been explicitly tested. Identifying the specific role played by each component of the model is difficult – as recognised by Sobal et al. (2006, p. 2), 'the components of the model ... are not mutually exclusive of each other because they overlap and interact'. Establishing causality is thus difficult. However, despite these limitations, it is important to recognise that qualitative research can play a useful role in understanding consumer motivations and mindsets in relation to food choice. For example, qualitative research has been fundamental to understanding the values that underpin the acceptance/rejection of organic food (Makatouni 2002).

2.4 Theory of Reasoned Action (TRA)/Theory of Planned Behaviour (TPB)

The theory of reasoned action (TRA), developed by Ajzen and Fishbein (1980) is premised on the belief that the immediate predictor of behaviour is a behavioural intention. Behavioural intentions depend on a person's attitude towards that behaviour and the values of others (Fig. 2.4). Attitudes capture beliefs about behavioural outcomes (behavioural beliefs) combined with an evaluation of the

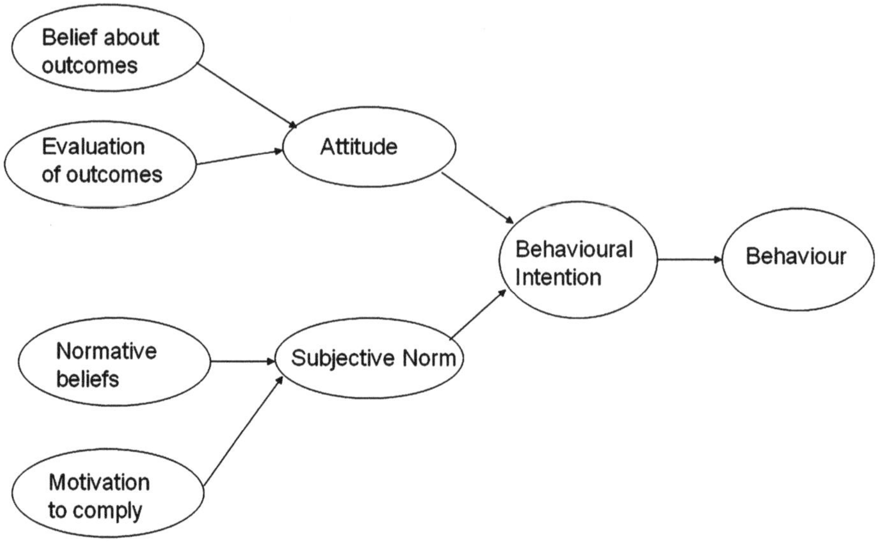

Fig. 2.4 Theory of reasoned action (TRA) (*Source*: adapted from Ajzen 2005, p. 126)

outcomes of such behaviour (outcome evaluation). Subjective norms refer to received social pressures to behave in a certain way and motivation to comply with the wishes of others.

The TRA was not specifically designed to model food choice, but it has been used extensively for this purpose. For instance, McCarthy et al. (2003) use the TRA to model beef consumption in Ireland. They found that both attitudes and subjective norms are important determinants of behavioural intentions, and that the latter were correlated significantly with their measure of behaviour. Other studies applying the TRA to food choice have also found significant correlations between the main components of the model (Saunders and Rahilly 1990).

However, the TRA was developed only to model purely volitional behaviour, in other words, cases where successful performance of the behaviour required only the formation of an intention to perform that behaviour (Conner and Armitage 2006). To address cases of incomplete volitional control, Ajzen (1991) developed the theory of planned behaviour (TPB). The TPB extends the TRA by incorporating an additional construct, perceived behavioural control (PBC), which is regarded as a determinant of both behavioural intensions and behaviour (Fig. 2.5). PBC incorporates both internal control factors (information, skills and abilities) and external control factors (dependence on others/situational factors). Therefore, according to the TPB, intentions are influenced by three factors: (a) whether the person is in favour of doing the specific behaviour (attitude towards the behaviour), (b) how much the person feels social pressure to do it (subjective norm), and (c) whether the person feels in control of the action in question or self-efficacy in relation to the behaviour (perceived behavioural control).

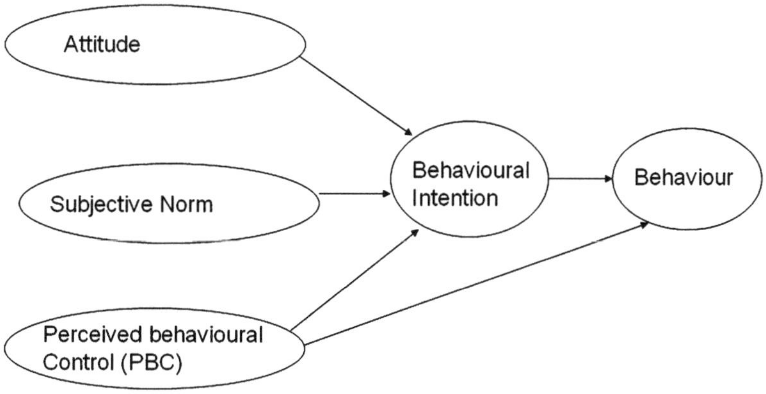

Fig. 2.5 Theory of planned behaviour (TPB) (*Source*: adapted from Ajzen 1991, p. 182)

It should be noted that the TPB still represents a 'reasoned action' approach to consumer behaviour because it assumes that intentions and behaviour in this domain follow reasonably from the behavioural, normative and control beliefs people hold about the behaviour. Although the beliefs people hold may be unfounded, inaccurate or even irrational, the attitudes, subjective norms and perceptions of behavioural control are thought to follow spontaneously and reasonably from these beliefs, produce a corresponding behavioural intention and ultimately result in behaviour that is consistent with the overall tenor of the beliefs. So by this theory, as a general rule, it is assumed that attitudes towards available options, which can be measured directly, are determinants of consumer decisions (Ajzen 1991).

The TPB has become the most widely adopted theoretical framework for modelling food choice (Conner and Armitage 2006). For instance, Cox et al. (1998) and Nguyen et al. (1996) employ the TPB to explain variations in fruit and vegetable and fat consumption, respectively. The TPB in Cox et al.'s (1998) study of fruit and vegetable consumption accounted for between 33 and 47% of the variation in behavioural intentions. Attitudes, subjective norms and PBC were all significant, with attitudes as the most important predictor. Nguyen et al. (1996) also found all three components to be significant, with attitudes as the most important factor in explaining intentions to eat fatty foods. Further studies (Povey et al. 2000) on fruit and vegetable consumption have found a significant linkage between behavioural intentions and actual behaviour. Given the large number of empirical studies that have validated the model specifically for the study of food choice, the TPB was selected as a core theoretical framework for the Focus Balkans project.

While numerous studies have demonstrated the validity of the TPB approach, several extensions have been proposed to improve the model fit. Two additional factors commonly included are self-identity and perceived need (Conner and Armitage 2006). Self-identity refers to the relatively enduring characteristics that people ascribe to themselves (Sparks 2000). In food choice, individuals may be more likely to eat healthily if they perceive themselves as 'health conscious' or eat environmentally friendly foods, such as organics, if they identify themselves as 'green consumers'. The latter notion was tested by Sparks and Shepherd (1992) who found that self-identity did significantly contribute to explaining intentions to consume organic vegetables, in addition to other components of the TPB model. A meta-analysis by Conner and Armitage (1998) suggests, however, that self-identity's contribution to the explanation of variations in behavioural intentions may be rather limited, certainly accounting for less than the 'traditional' components of the TPB.

Perceived need represents a second modification to the TPB, incorporated by Paisley and Sparks (1998). The latter argue that while the TPB may capture attitudes, it does not in itself assess whether individuals perceive a need for such behaviour. For example, an individual may regard a low-fat diet in general to be positive but perceive it as unnecessary to himself/herself. Studies which have incorporated perceived need have found that it adds significantly to the prediction of behavioural intentions (Paisley and Sparks 1998; Povey et al. 2000).

Attitude–intention relationships are likely to be weaker where individuals possess attitudinal ambivalence – simultaneously holding both positive and negative attitudes towards an object. Regarding food choice, for instance, an individual may hold both positive and negative attitudes to 'junk food', liking the taste but disliking the high-fat content. Conner and Armitage (2006, p. 52) argue that attitudinal ambivalence is likely to moderate the relationship between attitudes and intention/behaviour 'such that stronger (i.e. less ambivalent) attitudes are more predictive'. Sparks et al. (2001) found some empirical support for this proposition in studies focusing on eating meat and chocolate.

While researchers agree on the importance of attitudes in shaping behaviour, it is important to note that the model does not explain how attitudes are created or modified. The latter is particularly important for food agencies and commercial practitioners that wish to change behaviour. It may be that the formation of attitudes and their modification can only be explained in relation to the notions of the life course, trajectories and transitions introduced by Furst et al. (1996). Incorporating this into TPB modelling would, however, require a very different research design.

2.5 Conclusion

From reviewing the models, a number of conclusions can be drawn. First, explanations of cross-national variations in food choice cannot be solely reduced to differences in incomes. Culture plays an important part in explaining cross-national variations, but researchers have often struggled to capture this in their models. Second, past behaviour remains a consistent predictor of current behaviour. This is captured by Furst et al. (1996) in their notion of the life course and trajectories. Studies by social psychologists have found 'past behaviour predicts subsequent behaviour, over and above the effects of TPB variables' (Conner and Armitage 2008, p. 41). Given the strong linkage between past and current behaviour, studies of food choice should incorporate a historical dimension. Finally, while several studies applying the TPB indicate the significance of the main elements of the model and its overall validity, a thorough understanding of food choice requires also a consideration of how attitudes are generated and altered.

References

Ajzen I (1991) The theory of planned behaviour. Organ Behav Hum Decis Process 50:179–211

Ajzen I (2005) Attitudes, personality and behavior, 2nd edn. Open University Press, Maidenhead

Ajzen I, Fishbein M (1980) Understanding attitudes and predicting social behaviour. Prentice-Hall, Englewood Cliffs

Baltas G, Doyle P (2001) Random utility models in marketing research: a survey. J Bus Res 51(2):115–125

Becker G (1965) The theory of the allocation of time. Econ J 75:493–517

Bonke J (1992) Choice of foods: allocation of time and money, household production and market services. Copenhagen, MAPP working paper No. 3

Bove CF, Sobal J, Rauschenbach BS (2003) Food choices among newly married couples: convergence, conflict, individualism, and projects. Appetite 40(1):25–41

Cicia G, Del Giudice T, Scarpa R (2002) Consumers' perception of quality in organic food: a random utility model under preference heterogeneity and choice correlation from rank-orderings. Br Food J 104(3/4/5):200–213

Conner M, Armitage CJ (1998) Extending the theory of planned behavior: a review and avenues for further research. J Appl Soc Psychol 28:1429–1464

Conner M, Armitage CJ (2006) Social psychological models of food choice. In: Shepherd R, Raats M (eds) The psychology of food choice. CABI Publishing, Wallingford, pp 41–58

Conner M, Armitage CJ (2008) The social psychology of food. Open University Press, Maidenhead

Cox DN, Anderson AS, Lean MEJ, Mela D (1998) UK consumer attitudes, beliefs, and barriers to increasing fruit and vegetable consumption in the UK. Int J Public Health Nutr 1(1):61–68

Devine CM, Connors M, Bisogni CA, Sobal J (1998) Life-course influences on the development of a food choice trajectory: a qualitative analysis of fruit and vegetable use. J Nutr Educ 30:361–370

Falk LW, Bisogni CA, Sobal J (1996) Food choice processes of older adults: a qualitative investigation. J Nutr Educ 28(5):257–265

Falk LW, Bisogni CA, Sobal J (2000) Personal, social and situational influences associated with diet changes of participants in an intensive heart program. J Nutr Educ 32:251–260

Furst T, Connors M, Bisogni C, Sobal J, Falk L (1996) Food choice: a conceptual model of the process. Appetite 26(3):247–265

Gracia A, de Magistris T (2008) The demand for organic foods in the South of Italy: a discrete choice model. Food Policy 33(5):386–396

Green C (2008) Demand for allotments soars in quest for organic food. The Independent, 24th March, http://www.independent.co.uk/news/uk/this-britain/demand-for-allotments-soars-in-quest-for-organic-food-799872.html. Accessed 11 May 2012

Grunert KG, Baadsgaard A, Larsen HH, Madsen TK (1996) Market orientation in food and agriculture. Kluwer Academic Publishers, Dordrecht

Hanemann WE (1984) Discrete/continuous models of consumer demand. Econometrica 52(3):541–561

Lampietti JA, Lugg DA, Van Der Celen P, Branczik A (2009) The changing face of rural space: agriculture and rural development in the Western Balkans. World Bank, Washington, D.C

Makatouni A (2002) What motivates consumers to buy organic food in the UK?: results from a qualitative study. Br Food J 104(3/4/5):345–352

McCarthy M, de Boer M, O'Reilly S, Cotter L (2003) Factors influencing intention to purchase beef in the Irish market. Meat Sci 65(3):1071–1083

Nguyen MN, Otis J, Potvin L (1996) Determinants of intention to adopt a low-fat diet in men 30 to 60 years old: implications for heart health promotion. Am J Health Promot 10:201–207

Paisley CM, Sparks P (1998) Expectations of reducing fat intake: the role of perceived need within the theory of planned behaviour. Psychol Health 33:341–353

Povey R, Conner M, Sparks P, James R, Shepherd R (2000) Application of the theory of planned behaviour to two dietary behaviours: roles of perceived control and self-efficacy. Br J Health Psychol 5:121–139

Randall E, Sanjur D (1981) Food preferences – their conceptualization and relationship to consumption. Ecol Food Nutr 11(3):151–161

Saunders RP, Rahilly SA (1990) Influences on intention to reduce dietary intake of fat and sugar. J Nutr Educ 22:169–176

Schumacher SK, Boland MA (2005) The persistence of profitability among firms in the food economy. Am J Agric Econ 87(1):103–115

Sheth JN (2011) Models of buyer behavior: conceptual, quantitative and empirical. Marketing Classics Press, Decatur

Sobal J, Bisogni CA, Devine CM, Jastran M (2006) A conceptual model of the food choice process over the life course. In: Shepherd R, Raats M (eds) The psychology of food choice. CABI Publishing, Wallingford, pp 1–18

Sparks P (2000) Subjective expected utility -based attitude-behaviour models: the utility of self-identity. In: Terry DJ, Hogg MA (eds) Attitudes, behaviour and social context: the role of norms and group membership. Erlbaum, Mahwah, pp 31–46

Sparks P, Shepherd R (1992) Self-identity and the theory of planned behaviour: assessing the role of identification with green consumerism. Soc Psychol Q 55(4):388–399

Sparks P, Conner M, James R, Shepherd R, Povey R (2001) Ambivalence about health-related behaviours: an exploration in the domain of food choice. Br J Health Psychol 6:53–68

Swoboda B, Morschett D (2001) Convenience-oriented shopping. In: Frewer LJ, Risvik E, Schifferstein H (eds) Food, people and society. Springer, Heidelberg, pp 177–196

Chapter 3
Quantitative Surveys of Food Consumption and Motives: The Food Choice Questionnaire (FCQ)

Jasna Milošević Đorđević and Iris Žeželj

Abstract Data about the motives of consumers for buying and consuming specific food products are needed in order to tailor media messages and public health policies to the current nutritional challenges. Research in this field is especially important in the Western Balkan countries, where the population suffers from several diseases linked to unhealthy nutrition and the data on the actual eating habits are scarce. Addressing the motives behind different nutritional choices could help improve diets in the region. To this end, a Food Choice Questionnaire was applied to six Western Balkan countries. This method provided a systematic cross-national comparison of food choice motivations in the region as well as comparison to countries of the European Union.

Although the factor analysis resulted with eight, instead of nine, factors reported in the normative sample, the overall structure of Western Balkan food choice motives did not differ substantially from the initial study. The ranking of motives was proven to be fairly consistent across all Western Balkan states with 'sensory appeal' and 'purchase convenience' rated as the most important factors. Furthermore, this ranking is similar to the ranking obtained in other European countries with different traditions, cuisines and market conditions.

Keywords Food Choice Questionnaire • Western Balkans • Factor analysis • Food choice motives • Consumer behaviour

---

J.M. Đorđević, Ph.D. (✉)
Ipsos Strategic Marketing, Gavrila Principa 8, 11000 Belgrade, Serbia
e-mail: jasna.milosevic@ipsos.com

I. Žeželj, Ph.D.
Faculty of Philosophy, Belgrade University, Cika Ljubina 18-20, 11000 Belgrade, Serbia
e-mail: izezelj@f.bg.ac.rs

D. Barjolle et al. (eds.), *Food Consumer Science: Theories, Methods and Application to the Western Balkans*, DOI 10.1007/978-94-007-5946-6_3,

3.1 Introduction

Imagine you see a teenage girl buying a fruit salad in a cake shop. Why is she doing this? Maybe she likes fruits, maybe she is dieting, maybe all her friends buy healthy snacks and maybe she is fasting and is not allowed to eat dairy products. It is hypothesised that the relation between external influences and actual food choice behaviour is mediated by food-related attitudes and beliefs (Pettinger et al. 2004; Sun 2008). When making dietary choices, people perceive food as a means for satisfying different needs other than solely nutrition (e.g. appearance, lifestyle, image, healthiness). Bearing in mind that actual food consumption is affected by a multitude of influences, it is important to measure the relative salience of potential food choice motives in different populations. Measuring the motives behind food choice can help in tailoring media messages and health promotional campaigns to the needs of specific market segments. Studies prove that taste is still the single most important feature of food, which consumers are usually unwilling to compromise on (Verbeke 2006). However, other motives are growing in importance: a typical European food consumer is now looking for healthier and 'greener' (locally produced, less processed) alternatives (Grunert 2005). At the same time, the pace of life is dramatically changing, so the demand for food that is easily prepared is also on the rise. Studying the motive structure and its changes through survey-based research, particularly the implementation of the Food Choice Questionnaire (FCQ), is the focus of this chapter.

3.2 Method

3.2.1 Origin of Instruments Used

One of the most comprehensive instruments developed to assess both food- and non-food-related food choice motives is the Food Choice Questionnaire (FCQ). Steptoe et al. (1995) developed a protocol, the FCQ, which has been widely used to assess the impact of different motives on food choice. The authors initially identified nine factors that underpin food choice motivations: health, convenience (ease of preparation and availability), price, sensory appeal (appearance, taste and smell), mood, natural content (e.g. no additives), weight control (low in calories and fat), familiarity and ethical concern (politically approved country of origin, environmentally friendly packaging). This protocol was successfully implemented as a whole or partially in urban English-speaking populations (Glanz et al. 1998; Lockie et al. 2002; Martins and Pliner 1998; Pollard et al. 1998) and, more recently, in urban non-English-speaking populations (Ares and Gambaro 2007; Biloukha and Utermohlen 2000; Fotopoulos et al. 2009; Honkanen and Frewer 2009; Januszewska et al. 2011; Lindeman and Väänänen 2000). The initial nine factorial structures, proposed by Steptoe et al. (1995), did not prove to be invariant across Western European populations (Eertmans et al. 2005; Fotopoulos et al. 2009).

3.2.2 Searching for Underlying Constructs: Factor Analysis

The aim of factor analysis is to reveal latent variables that cause the covariation of manifest variables. There are two main types of factor analysis: exploratory (EFA) and confirmatory (CFA) that serve different roles – to explore and identify or to confirm already existing assumptions. EFA is used to uncover the underlying structure of a relatively large set of variables, while CFA seeks to confirm expected relations on the basis of previous research or theory. While EFA is oriented towards discovering the model, confirmatory analysis is a deductive method that tests specific hypothesis. EFA is most suitable to explore the structure of the data describing variability among observed variables and to search for a potentially lower number of unobserved factors. The observed variables are modelled as linear combinations of the potential factors, thus resulting in a smaller number of underlying constructs. Highly intercorrelated variables are grouped into a distinct set of groups by EFA, and the researcher interprets the interrelationships between the factors.

In contrast to traditional EFA that does not focus on the structure of the model being tested, CFA requires the researcher to specify both the number of factors and patterns of variable loadings on underlying factors (Hair et al. 2010; Little 2000; Matsumoto and Hee Yoo 2006; Cheung and Leung 2006). In applying the method and meeting the objectives of factor analysis, the researcher has to specify the research question, obtain the data and consider conceptually the structure of the data, decide which type of factor analysis needs to be applied and, after the application, interpret the results. Sample size should be adequate: at least 50–100 cases (Barrett and Kline 1981; Comfrey and Lee 1992) or with a subject to item ratio of at least 10:1 (Nunnally 1978, p. 421).

With the aim of assessing the structure of motives underpinning food choice in Western Balkan countries (WBC), we conducted a factor analysis. Given the existence of previous research, as a first step, we estimated a fit of Steptoe et al.'s (1995) nine independent-factor FCQ model for each country as well as the combined WBC sample. Goodness of fit was assessed using a number of measures. These included the chi-square test as a descriptive goodness-of-fit index for nested models, the comparative fit index (CFI), the parsimonious comparative fit index (PCFI) and the root mean square error of approximation (RMSEA). As the nine-factor model demonstrated suboptimal fit, we turned to EFA, in order to reveal the factor structure of the instrument for this specific population (Fotopoulos et al. 2009).

3.2.3 Protocol and Data Collection

Implementation of the FCQ followed three stages and covered six countries: Bosnia and Herzegovina (BiH), Croatia, the Former Yugoslav Republic of Macedonia (FYROM), Montenegro, Serbia and Slovenia. The first stage was translating the FCQ from English into local languages. The questionnaire was translated first into Serbian and back translated to English by a second translator, not familiar with the instrument (as suggested by Brislin 1986). The original English version and the back-translated version were then matched by two independent observers. Minor discrepancies were corrected until the back-translated version fully matched the original English version. Based on the Serbian questionnaire, five separate versions were developed for the other studied countries in the respective national language. One difference from the original FCQ is that we opted for a five-point instead of the original four-point scale used by Steptoe et al. (1995). This was for two main reasons: (a) in order to avoid a forced (artificial) agreement or disagreement of respondents and (b) for the scale range to match others in the questionnaire (Cox 1980; Saris and Gallhofer 2007).

In the second stage of implementation, a pilot survey was conducted. This was in order to assess the clarity and relevance of each item in every country. It included 60 respondents (10 per country). Feedback from the pilot survey led to the wording of two FCQ items being modified to improve clarity. Overall, however, the pilot study confirmed the appropriateness of the instrument and only minor changes were introduced.

The final phase of implementation focused on sampling and data collection. Data collection for the main sample occurred via face-to-face interviews, conducted in the respondents' homes. In this study the sampling universe was based on census data for each country and estimated population dynamics. In each country the sample was nationally representative, and a stratified three-stage random sampling approach was adopted. The three stages were as follows. Primary sampling units were polling station territories (PSUs) in all countries except in Slovenia where enumeration areas were used as approximate PSUs. PSUs enable the most reliable sampling, due to the fact that these units provide the most complete data in the countries studied (register of dwelling addresses). Each polling station territory comprised approximately 200 households and was defined by street(s) name(s). Secondary sampling units were households: one household comprised people living in the same dwelling. Tertiary sampling units were actual respondents. In order to optimise the sample plan and reduce sampling error, the stratification for each country was done by region and type of settlement. In each country the sample size was around 500, giving a total sample of 3,085 responses.

3.3 Scopes and Objectives of the FCQ Study

The 2004 and 2007 enlargements of the European Union (EU) were historic achievements: the accession of ten countries from Central and Eastern Europe with radically different socioeconomic and administrative histories compared with established member states. In the short to medium term, the main focus of EU accession debates will now focus on the so-called Western Balkan countries. It is of utter importance to obtain detailed baseline data on these countries for considering the current national situation and the potential impact of EU accession and for crafting relevant policy responses. Research in the field of food consumer science is particularly important in the WBC, where the population suffers from several diseases linked with insufficient nutrition in quantity but also in quality.

Previous research suggested that a high proportion of the population in the WBC has an unbalanced diet, which places them at a high risk of different diseases, especially coronary (Kromhout 2001; Menotti et al. 1999). It is therefore desirable to improve diets in the region. This, however, requires a detailed understanding of how consumers choose between food products and motivations underlying such choices.

The objective of this research was to analyse the motives for food choice, using the FCQ in six WBC. The research provided a systematic cross-national comparison of food choice motivations in the WBC, assessing the generalizability of the FCQ factorial structure to a markedly different environment from the original United Kingdom sample and other Western European countries (which account for the majority of applications of the FCQ).

3.4 Results

CFA was used to measure the fit with the original model of nine underlying independent motives for food choice as proposed by Steptoe et al. (1995). We tried to determine whether the constructs are distinct from each other and stable across the six countries. The analysis was performed for each country separately as well as for the pooled sample. In all cases, the level of fit between the data for the WBC and the original nine independent dimensions was poor (see CFA1 parameters in Table 3.1). Table 3.1 displays the goodness-of-fit indices: chi-square (the normal theory maximum likelihood chi-square), CFI (comparative fit index) and RMSEA (root mean square error of approximation). A highly significant chi-square shows that a significant amount of observed covariance between items remains unexplained by the original model (considering the large sample size in this study, chi-square indices must be interpreted carefully). Other measures of fit also lie outside of conventional acceptance limits. RMSEA values were above the acceptance limit, and the CFI scores are below the recommended level.

A RMSEA value of 0.05 indicates a close fit and values up to 0.08 represent reasonable errors of approximation in the population (Browne and Cudeck 1993). In our case, CFA parameters suggest that the model cannot be accepted ($\chi^2 = 18{,}033$; df = 599; CFI = 0.543; RMSEA = 0.097). The best fit of the model is for the Slovenian and the total WBC sample.

The goodness-of-fit summary statistics indicate, therefore, that the original nine-factor structure proposed by Steptoe et al. (1995) does not provide an optimal fit for any of the WBC countries. Although the Steptoe et al. (1995) study is more than 15 years old, more recent data suggest that for some samples the nine-factor structure still holds (Januszewska et al. 2011), so the results for the WBC cannot be attributed solely to changes in the global market environment. They might also be due to the specific characteristics of food consumers in the region. In order to investigate the scope of these discrepancies, in the next stage we conducted an EFA.

The EFA for the pooled WBC sample extracted eight factors instead of the original nine. When comparing the WBC results with the original factors identified from Steptoe et al.'s (1995) instrument, it appears that although the factor structure is

Table 3.1 Confirmatory factor analysis (parameters of fit)

	Total sample	Serbia	FRYOM	Montenegro	BIH	Croatia	Slovenia
Chi-square	18033.398	4406.289	3859.849	4185.005	4061.026	3986.913	3139.792
Degrees of freedom	599	599	599	599	599	599	599
Probability level	0.000	0.000	0.000	0.000	0.000	0.000	0.000
CFI	0.543	0.515	0.456	0.462	0.502	0.529	0.565
PCFI	0.516	0.489	0.434	0.440	0.477	0.503	0.537
RMSEA	0.097	0.111	0.103	0.106	0.106	0.106	0.092

Source: Reprinted from Miloševi et al. (2012, p. 208). With permission from Elsevier

similar, there are some significant differences (Table 3.2). Most of the items from the original factor structure demonstrate high (>0.70) or moderate (>0.40) loadings on original factors and loading on other factors (<0.50). However, a few items cross-load (>0.50) on other factors (see third column of Table 3.2). Specifically, item 2 (*Contains no additives*) from the 'natural content' factor loads onto 'weight control', while item 12 (*Is good value for money*) from the 'price' factor loads onto the 'purchase

Table 3.2 Results from the exploratory factor analysis, factor loadings on numbered items

Item number	Original factors	Extracted factors	Item loading >0.50 on other factors besides original
	1. Health	1 and 5 Health, natural content	
22	Contains a lot of vitamins and minerals	0.65	
29	Keeps me healthy	0.69	
10	Is nutritious	0.40	
27	Is high in protein	0.45	
30	Is good for my skin/ teeth/hair/nails	0.62	
9	Is high in fibre and roughage	0.31	
	2. Mood	2. Mood	
16	Helps me cope with stress	0.71	
34	Helps me to cope with life	0.58	
26	Helps me relax	0.70	
24	Keeps me awake/ alert	0.38	Health 0.55
13	Cheers me up	0.64	
31	Makes me feel good	0.47	Health 0.55
	3. Convenience	3.1. Preparation convenience	3.2. Purchase convenience
1	Is easy to prepare	0.81	
15	Can be cooked very simply	0.74	
28	Takes no time to prepare	0.73	
35	Can be bought in shops close to where I live or work		0.47
11	Is easily available in shops and supermarkets		0.71
	4. Sensory appeal	4. Sensory appeal	

(continued)

Table 3.2 (continued)

Item number	Original factors	Extracted factors	Item loading >0.50 on other factors besides original
14	Smells nice	0.69	
25	Looks nice	0.44	
18	Has a pleasant texture	0.62	
4	Tastes good	0.60	
	5. Natural content	1 and 5 Health, natural content	
2	Contains no additives	0.32	Weight control 0.51
5	Contains natural ingredients	0.55	
23	Contains no artificial ingredients	0.62	
	6. Price	6. Price	
6	Is not expensive	0.76	
36	Is cheap	0.77	
12	Is good value for money	0.17	Purchase convenience 0.61
	7. Weight control	7. Weight control	
3	Is low in calories	0.75	
17	Helps me control my weight	0.51	
7	Is low in fat	0.74	
	8. Familiarity	8 and 9 Familiarity, ethical concern	
33	Is what I usually eat	0.47	
8	Is familiar	0.31	
21	Is like the food I ate when I was a child	0.65	
	9. Ethical concern	8 and 9 Familiarity, ethical concern	
20	Comes from countries I approve of politically	0.71	
32	Has the country of origin clearly marked	0.58	
19	Is packaged in an environmentally friendly way	0.55	

Source: Reprinted from Milošević et al. (2012, p. 209). With permission from Elsevier

convenience' factor. Items 24 (*Keeps me awake/alert*) and 31 (*Makes me feel good*) from the 'mood' factor load onto the 'health and natural content' factor.

Although Steptoe et al. (1995) extracted one factor that comprised purchase and preparation convenience, our data suggested that these are perceived as independent factors. Convenience splits into two factors: purchase convenience (items 11, 12 and 35) and preparation convenience (items 1, 15 and 28). The FCQ items from the 'health' and 'natural content' scales, on contrast, loaded onto a single factor. The same occurred with 'familiarity' and 'ethical concern'.

Some of these discrepancies from the original normative sample are similar to differences uncovered by others. For instance, a single 'health factor', containing items relating to health benefits and natural characteristics (absence of artificial ingredients or additives), was also obtained in research for Belgium and Italy (Eertmans et al. 2006), suggesting that food consumers view both safety and the nutritious value of the food to be essential for its healthiness. Russian food consumers (Honkanen and Frewer 2009) also viewed preparation convenience to be different from purchase convenience (availability), with higher importance attributed to the latter motive.

The elaborated differences, however, do not allow us to conclude that there is a substantial change in motivational typology of food choice between the normative sample and the WBC sample. Although EFA yielded a different factorial structure, the basic motives proposed by Steptoe et al. (1995) still emerged.

As detailed in Table 3.3, the reliability of the overall FCQ typology in our study is very high (Cronbach's alphas=0.920). None of the new subscales possessed a Cronbach's alpha lower than 0.60. The average importance given to the 36 FCQ items is 3.94, on the 1–5 scale. The highest importance of individual items is given to the following: *tastes good* (4.60), *keeps me healthy* (4.44), *contains natural ingredients*

Table 3.3 Descriptive statistics and reliabilities of the original FCQ (N=3,085)

Item number	It is important to me that the food I eat on a typical day:	Mean	SD
	1. Health and natural content		
22	Contains a lot of vitamins and minerals	4.29	0.86
29	Keeps me healthy	4.44	0.77
10	Is nutritious	4.34	0.79
27	Is high in protein	3.86	1.00
30	Is good for my skin/teeth/hair/nails etc.	4.17	0.98
9	Is high in fibre and roughage	3.81	1.01
5	Contains natural ingredients	4.40	0.79
23	Contains no artificial ingredients	4.15	0.99
24	Keeps me awake/alert	4.10	0.97
31	Makes me feel good	4.21	0.89
	Cronbach's alpha: 0.85, mean: 4.18		
	2. Mood		
16	Helps me cope with stress	3.52	1.19
34	Helps me to cope with life	3.75	1.11

(continued)

Table 3.3 (continued)

Item number	It is important to me that the food I eat on a typical day:	Mean	SD
26	Helps me relax	3.76	1.09
13	Cheers me up	3.87	1.07
	Cronbach's alpha: 0.800, mean: 3.72		
	3. Preparation convenience		
1	Is easy to prepare	3.47	1.18
15	Can be cooked very simply	3.98	1.07
28	Takes no time to prepare	3.74	1.15
	Cronbach's alpha: 0.74, mean: 3.85		
	4. Purchase convenience		
35	Can be bought in shops close to where I live or work	4.07	0.95
11	Is easily available in shops and supermarkets	4.21	0.88
12	Is good value for money	4.39	0.82
	Cronbach's alpha: 0.63, mean: 4.22		
	5. Sensory appeal		
14	Smells nice	4.39	0.79
25	Looks nice	4.02	1.00
18	Has a pleasant texture	4.15	0.94
4	Tastes good	4.60	0.63
	Cronbach's alpha: 0.65, mean: 4.29		
	6. Price		
6	Is not expensive	4.11	0.98
36	Is cheap	4.00	1.05
	Cronbach's alpha: 0.72, mean: 4.05		
	7. Weight control		
3	Is low in calories	3.47	1.18
17	Helps me control my weight	3.64	1.18
7	Is low in fat	3.86	1.04
2	Contains no additives	3.90	1.07
	Cronbach's alpha: 0.71, mean: 3.72		
	8. Familiarity and ethical concern		
33	Is what I usually eat	3.92	1.00
8	Is familiar	4.02	0.99
21	Is like the food I ate when I was a child	3.49	1.28
20	Comes from countries I approve of politically	2.53	1.36
32	Has the country of origin clearly marked	3.51	1.26
19	Is packaged in an environmentally friendly way	3.35	1.24
	Cronbach's alpha: 0.71, mean: 3.47		

Source: Reprinted from Milošević et al. (2012, p. 209). With permission from Elsevier
Note: Five-point Likert scale was used

(4.40), *smells nice* (4.39), *is good value for money* (4.39) and *is nutritious* (4.33). An examination of variance reveals that the diversity of answers is highest on 'ethical concern', followed by the 'familiarity' and 'weight control' subscales, while less dispersion from the mean is detected for those factors ranked in general as more important.

For the sample as a whole, the highest importance of individual subscales measured through mean scores is recorded for 'sensory appeal' (4.29) and then for 'purchase convenience' (4.22), 'health and natural content' (4.18) and 'price' (4.05). The lowest importance is assigned to 'weight control' (3.72) and 'familiarity and ethical concern' (3.47). None of the factors in general thus appears irrelevant in motivating food choice.

3.5 Discussion

A key research question in cross-cultural studies is whether psychological processes are universal across countries, which could be confirmed by construct comparability (factorial invariance) for different countries. We tried to establish as much equivalence as possible in all aspects of the research process to allow cross-national and cross-cultural comparison: from the sampling procedure, translation and data collection to the applied analysis of the instruments. Although Steptoe et al.'s (1995) original structure of nine independent food choice motives has not been confirmed by CFA in this study, the structure of WBC food choice motives did not differ essentially from the initial study, conducted 20 years ago for a different market.

The cross-national analysis revealed far more similarities than differences between the WBC, so it can be argued that the region can be treated as a fairly homogeneous market, when it comes to food choice motives and nutrition habits.

The ranking of motives for the WBC mainly reaffirms studies in other European countries (Table 3.4), as the same factors are reported to be the most important in British (Steptoe et al. 1995; Pollard et al. 1998), Belgian, Italian (Eertmans et al. 2005), Russian (Honkanen and Frewer 2009), Finnish (Lindeman and Väänänen 2000) and Greek samples (Fotopoulos et al. 2009). The highest rating for 'sensory appeal' mirrors Steptoe et al.'s (1995) results and the findings of Januszewska et al. (2011) for Hungary and Romania. In other European studies, similarly 'ethical concern' and 'familiarity' were found to be the least important motives for dietary choice.

3.6 Conclusions

This research presents the first cross-national assessment of food choice motives in the WBCs. Overall, the ranking of motives was proven to be fairly consistent across all WBC states with 'sensory appeal', 'purchase convenience' and 'health and natural content' rated as the most important factors and 'familiarity and ethical concerns' rated as least important. Furthermore, this ranking is strikingly similar to the ranking obtained in other European countries with different traditions, cuisine and market conditions.

Table 3.4 Mean values for food choice motives in EU countries

Study	Milosevic et al. (2012)	Lindman and Vaananen (2000)	Eertmans et al. (2006)		Fotopoulos et al. (2009)	Honkanen and Frewer (2009)	Steptoe et al. (1995)	Januszewska et al. (2011)	
Country	WBC	Finland	Italy	Belgium	Greece	Russia	Great Britain	Romania	Hungary
Factor in order of importance	1–5	1–4	1–4	1–4	1–7	1–7	1–4	1–7	1–7
Most important	Sensory appeal (4.3)	Price (2.8)	Natural content (3.2)	Sensory appeal (2.9)	Natural content (5.9)	Sensory appeal (6.2)	Sensory appeal (3.0)	Sensory appeal	Sensory appeal
Second	Purchase convenience (4.2)	Health (2.8)	Health (3.1)	Price (2.8)	Health (5.6)	Natural content (5.6)	Health (2.8)	Natural content	Natural content
Third	Health and natural content (4.2)	Sensory appeal (2.8)	Sensory appeal (3.1)	Health (2.6)	Sensory appeal (5.6)	Price (5.6)	Price (2.8)	Health	Convenience
Fourth	Price (4.0)	Convenience (2.6)	Price (2.7)	Convenience (2.5)	Price (5.6)	Health (5.5)	Convenience (2.8)	Price	Price
Fifth	Preparation convenience (3.8)	Ecological welfare (2.4)	Weight control (2.7)	Natural content (2.2)	Mood (5.3)	Mood (5.5)	Natural content (2.5)	Weight control	Health
Sixth	Mood (3.7)	Mood (2.4)	Convenience (2.7)	Weight control (2.1)	Convenience (5.3)	Familiarity (5.2)	Weight control (2.4)	Mood	Weight control
Seventh	Weight control (3.7)	Weight control (2.3)	Mood (2.6)	Mood (2.0)	Weight control (5.2)	Convenience (5.0)	Mood (2.1)	Convenience	Mood

Eighth	Familiarity and ethical concern (3.5)	Natural content (2.2)	Ethical concern (2.3)	Familiarity (1.9)	Ethical concern (5.2)	Ecological welfare (4.8)	Ethical concern (1.9)	Ethical concern	Familiarity
Nineth		Familiarity (1.8)	Familiarity (2.1)	Ethical concern (1.7)	Familiarity (5.0)	Weight control (4.2)	Familiarity (1.8)	Familiarity	Ethical concern
Tenth		Religion (1.7)				Political values (4.0)			
Eleventh		Political values (1.7)				Religion (3.2)			

The FCQ demonstrated satisfactory cultural invariance with respect to (a) factor configuration (although the match was not optimal, the differences in comparison to a normative study were not substantial), which indicates that the FCQ measures similar constructs across countries and (b) factor loadings (similar items were grouped together forming one underlying factor). These results are in line with recent cross-cultural studies (Januszewska et al. 2011; Pieniak et al. 2009) and do not support the conclusion of Fotopoulos et al. (2009) that a more basic, revised motivational typology of food choice is needed.

Comprehensive investigation of consumers' food choice motives is just the first step in understanding their actual food purchase and consumption behaviour. Cultural norms and broader situational factors determining food choice in the WBC were only lightly tackled in this study and could be the focus of further research. Future research on Western Balkan consumers could also explore their awareness of food safety issues and their perception of the importance of food safety. Detailed knowledge on so-called food system – economic, cultural and individual factors underpinning food choice – could help improve public communication campaigns that seek to promote healthier diets.

References

Ares G, Gambaro A (2007) Influence of gender, age and motives underlying food choice on perceived healthiness and willingness to try functional foods. Appetite 49:148–158

Barrett PT, Kline P (1981) The observation to variable ratio in factor analysis. Pers Study Group Behav 1:23–33

Biloukha O, Utermohlen V (2000) Correlates of food consumption and perceptions of food in an educated urban population in Ukraine. Food Qual Prefer 11:475–485

Brislin RW (1986) The wording and translation of research instrument. In: Lonner WJ, Berry JW (eds) Field methods in cross-cultural research. Sage, Beverly Hills, pp 137–164

Browne MW, Cudeck R (1993) Alternative ways of assessing model fit. In: Bollen KA, Long JS (eds) Testing structural equation models. Sage, Newbury Park, pp 136–162

Cheung M, Leung K (2006) Evaluating multilevel models in cross-cultural research. J CrossCult Psychol 37:522–541

Comfrey AL, Lee HB (1992) A first course in factor analysis. Lawrence Erlbaum Associates, Hillsdale

Cox EP (1980) The optimal number of response alternatives for a scale: a review. J Mark Res 27(4):407–422

Eertmans A, Victoir A, Vansant G, Van den Bergh O (2005) Food-related personality traits, food choice motives and food intake: mediator and moderator relationships. Food Qual Prefer 16:714–726

Eertmans A, Victoir A, Notelaers G, Vansant G, Van den Bergh O (2006) The food choice questionnaire: factorial invariant over western urban populations? Food Qual Prefer 17:344–352

Fotopoulos C, Krystallis A, Vassallo M, Pagiaslis A (2009) Food Choice Questionnaire (FCQ) revisited. Suggestions for the development of an enhanced general food motivation model. Appetite 52:199–208

Glanz K, Basil M, Maibach E, Goldberg J, Snyder D (1998) Why Americans eat what they do: taste, nutrition, cost, convenience and weight control concerns as influences of food consumption. J Am Diet Assoc 98:1118–1126

Grunert KG (2005) Food quality and safety: consumer perception and demand. Eur Rev Agric Econ 32:369–391

Hair J, Black W, Babin B, Anderson R (2010) Multivariate data analysis. Pearson Prentice Hall, Upper Saddle River

Honkanen P, Frewer L (2009) Russian consumers' motives for food choice. Appetite 52:363–371

Januszewska R, Pieniak Z, Verbeke W (2011) Food choice questionnaire revisited in four countries. Does it still measure the same? Appetite 57:94–98

Kromhout D (2001) Epidemiology of cardiovascular diseases in Europe. Public Health Nutr 4:441–457

Lindeman M, Väänänen M (2000) Measurement of ethical food choice models. Appetite 34:55–59

Little T (2000) On the comparability of constructs in cross-cultural research. J Cross Cult Psychol 31:213–219

Lockie S, Lyons K, Lawrence G, Mummery K (2002) Eating 'Green': motivations behind organic food consumption in Australia. Sociol Rural 42:23–40

Martins Y, Pliner P (1998) The development of the food motivation scale. Appetite 30:94

Matsumoto D, Hee Yoo S (2006) Toward a new generation of cross-cultural research. Perspect Psychol Sci 1(3):234–250

Menotti A, Kromhout D, Blackburn H, Fidanza F, Buzina R, Nissinen A (1999) Food intake patterns and 25-year mortality from coronary heart disease: cross-cultural correlations in the Seven Countries Study. Eur J Epidemiol 15:507–515

Milošević J, Žeželj I, Gorton M, Barjolle D (2012) Understanding the motives for food choice in Western Balkan countries. Appetite 58:205–214

Nunnally JC (1978) Psychometric theory, 2nd edn. McGraw Hill, New York

Pettinger C, Holdsworth M, Gerber M (2004) Psycho-social influences on food choice in Southern France and Central England. Appetite 42:307–316

Pieniak Z, Verbeke W, Vanhonacker F, Guerrero L, Hersleth M (2009) Association between traditional food consumption and motives for food choice in six European countries. Appetite 53:101–108

Pollard T, Steptoe A, Wardle J (1998) Motives underlying healthy eating: using the food choice questionnaire to explain variation in dietary intake. J Biosoc Sci 30:165–179

Saris WE, Gallhofer IN (2007) Design, evaluation and analysis of questionnaires for survey research. Wiley, Hoboken

Steptoe A, Pollard T, Wardle J (1995) Development of a measure of the motives underlying the selection of food: the food choice questionnaire. Appetite 25:267–284

Sun Y-HC (2008) Health concern, food choice motives, and attitudes toward healthy eating. The mediating role of food choice motives. Appetite 51:42–49

Verbeke W (2006) Functional foods: consumer willingness to compromise on taste for health? Food Qual Prefer 17:126–131

Chapter 4
Segmenting Consumers Using Cluster Analysis: An Application to Food Motivations in the Western Balkan Countries

Matthew Gorton, Mitchell Ness, and John White

Abstract To illustrate the use of cluster analysis as a tool for segmenting consumers, the Focus Balkans project applied this method to the topic of food motivations. The aim of cluster analysis is to group objects (e.g. people or products) on the basis of numerical measures. Food motivations were captured via the Food Choice Questionnaire. Respondents were clustered based on the eight factors identified in the previous chapter. The profiling of clusters allowed for the identification of groups that from a public health perspective raise concern. For the Western Balkans, five groups of consumers were discerned with the clusters characterised by significant differences in food consumption patterns as well as demographic and socioeconomic characteristics. Three clusters raise concern. For each cluster, the communication messages should be distinctive, and they are described briefly in this chapter.

Keywords Cluster analysis • Food motivations • Public health • Western Balkan countries

4.1 Introduction

The aim of cluster analysis (CA) is to group objects (e.g. people or products) on the basis of numerical measures. Objects are grouped (clustered) on the basis of similarity to one another. Objects within a group should be as 'similar' to one another as

M. Gorton, Ph.D. (✉) • M. Ness, Ph.D.
Marketing Subject Group, Newcastle University Business School,
5 Barrack Road, Newcastle upon Tyne NE1 4SE, UK
e-mail: matthew.gorton@newcastle.ac.uk; mitchell.ness@newcastle.ac.uk

J. White, Ph.D.
School of Management, Plymouth University,
Drake Circus, Plymouth, Devon PL4 8AA, UK
e-mail: john.white@plymouth.ac.uk

D. Barjolle et al. (eds.), *Food Consumer Science: Theories, Methods and Application to the Western Balkans*, DOI 10.1007/978-94-007-5946-6_4,

possible, whilst objects belonging to other groups should be as 'dissimilar' as possible (Everitt et al. 2011; Romesburg 1984).

CA can be a highly useful tool for marketers as it recognises the potential existence of segments, helps in their identification and then provides decision makers with insights that they can use to inform and improve decision making. As Saunders (1980, p. 442) notes, 'the groupings of similar objects provide a series of platforms between the two extreme views that: (a) all objects are unique and inviolable and (b) the population is homogeneous'. This chapter outlines the utility of cluster analysis and the main elements of the approach and details an application of the method. It is designed as an introduction to the topic for non-specialists, with an example indicating how it may be used by consumer food scientists.

4.2 Method

CA has been used extensively for segmentation in the field of consumer food science (e.g. Askegaard and Madsen 1995; Geeroms et al. 2008; Heinemann et al. 2006; Mai and Ness 1997; Mesias et al. 2003; Murphy et al. 2000; Saba and Messina 2003; Schnettler et al. 2009). For instance, Heinemann et al. (2006) applied CA to ascertain attitudes towards parboiled rice. Three clusters were identified. The group that displayed a negative attitude to parboiled rice (and therefore needs to be convinced of the merits of it) was found to have the highest income and education level and the highest frequency of consumption of milled rice. Subsequent analysis suggested that parboiled rice is negatively perceived, and the challenge for increasing demand is one of changing this perception through emphasis of the convenience and health benefits of consumption of parboiled rice.

CA can also be used as a tool for improving the targeting of marketing communications relating to public health initiatives. For example, Kennedy et al. (2005) used principal component analysis and hierarchical cluster analysis to group consumers based on their understanding and practice of food safety. The resultant analysis identified three food handler types: conscientious, cavalier and careful. Food preparers with the poorest food practices were typically younger (under 45 years of age), urban based, male and highly educated. Segmenting the groups in this way clearly enables policymakers to develop more informed and targeted food safety programmes.

In applying the method, the three main elements of the analysis are deciding the number of clusters (groups), determining membership of each group and profiling the characteristics of each group. CA employs measures of closeness of the objects using similarity measures for nonmetric data and distance measures for metric data. The technique requires measurement of distances between objects and between groups.

There are two main types of CA: hierarchical and optimisation methods. Hierarchical clustering begins with as many clusters as there are objects. On completion, there is a single cluster of all objects. The procedure employs information in a distance matrix, and at each stage a merger takes place so that the number of clusters is reduced by one. There are many specific hierarchical techniques, for instance, single link, complete link, centroid, median, Ward's and group average. Differences between these methods arise in the way similarity or distance between objects or groups is represented although the basic principle is the same.

The process is summarised by the agglomeration schedule, dendrogram and Gower diagram. The researcher uses these summary methods to decide the appropriate number of clusters. This involves the researcher's judgement in assessing the overall fit obtained within each grouping and the improvement that is obtained in this fit with the inclusion of an additional group.

The optimisation method groups objects into a prespecified number of clusters relative to an objective. It involves two stages in which there is an initial grouping of data followed by the application of a clustering criterion to reach a final solution. Typically, the clustering criterion is based on the relationship between total variance of the ungrouped objects and within and between group variance. Since total variance is fixed, the criterion either minimises within group variance or maximises between group variance.

A common approach is to combine hierarchical and optimisation methods. For instance, the hierarchical approach is used in the first stage to identify outliers, define the number of clusters and profile the cluster centres. In the second stage, an optimisation method is applied, using the cluster centres from the hierarchical results used as the initial seed points, to determine final cluster membership. The two-stage approach combines the advantages associated with hierarchical and non-hierarchical methods, whilst at the same time minimising the drawbacks linked to utilising each independently (Milligan and Cooper 1985).

Profiles of clusters are established based on a descriptive analysis of variables used in the analysis as well as other variables (demographic, behavioural and attitudinal) to establish significant differences between clusters. This is a critical part of the validation process. In the case of nominal profile variables, a Chi-square contingency test is used under the null hypothesis that cluster identity and the profile variable are independent. In the case of scale profile variables, an analysis of variance (ANOVA) test for the comparison of average cluster scores is used under the null hypotheses that the average cluster scores are equal.

In consumer research, however, the utility of CA may be limited by a number of factors. First, there are no 'dependent' and 'independent' variables specified, so that CA fails to provide the basis for ascertaining the determinants of outcomes. It is thus a tool for classification or dissection rather than establishing causal relationships. Second, researchers often depend on data for a single time period, making it difficult to assess the stability of the clusters identified. This weakens the usefulness of CA as an aid for predicting market developments and supporting decision-making. Finally, as with other techniques, CA will be limited where it fails to incorporate relevant or includes irrelevant variables (Punj and Stewart 1983). In this case, CA may lead to recommendations that fail to capture important motivations. There must be a clear rationale and theoretical basis for the inclusion of variables.

4.3 Context of the Study

To illustrate the use of cluster analysis as a tool for segmenting consumers, we extend the analysis presented in the previous chapter. The purpose of CA within the Focus Balkans project was to better understand food motivations, providing the basis for more targeted and hence effective communication messages and policy recommendations. Specifically, we segment consumers based on the factor analysis of responses to the Food Choice Questionnaire (FCQ). This recognises that factors motivating food choice are likely to vary significantly across consumer segments in the region. The FCQ is introduced in the previous chapter along with a discussion of the survey and sampling methods. A more detailed discussion of methods is presented in Milošević et al. (2012).

The cluster analysis followed a two-stage design. First, hierarchical cluster analysis was employed to define the number of clusters, identify outliers and profile the cluster centres. Second, an optimisation approach was used to determine final cluster membership using the centroids from the first stage as initial seed points.

4.4 Results

Respondents were clustered based on the eight factors identified in the previous chapter (health and natural content, mood, preparation convenience, purchase convenience, sensory appeal, price, weight control, familiarity and ethical concern). An analysis of the agglomeration schedule, dendrogram and Gower diagram indicated that a five-cluster solution was most appropriate. After outliers and records with missing data were removed, 2,814 responses were retained in the CA. The validation of the clusters depended on a set of additional variables relating to food consumption, demographic and socioeconomic characteristics.

Table 4.1 presents the average factor scores for the five clusters, whilst Table 4.2 allows for a comparison of clusters based on variables not included in the initial factor analysis. Table 4.3 details cluster membership by country. The remainder of this section profiles the clusters.

4.4.1 Cluster 1: 'Food Enthusiasts'

This cluster records above average mean scores for all factors. Compared to all the other clusters, Cluster 1 places greatest emphasis on 'sensory appeal' (tastes good, looks and smells nice, etc.) and 'preparation convenience'. This group also, compared to other clusters registers the highest mean score for the factor 'weight control'. This cluster has the highest proportion of females (59.7%). The mean income of this cluster is similar to the sample average. The average self-reported

Table 4.1 Average factor scores by cluster

	Cluster				
	1	2	3	4	5
	Food enthusiasts	Unconcerned	Price/distress	Purchase convenience	Health oriented
	$n=648$	$n=626$	$n=590$	$n=485$	$n=464$
Factor 1 (health)	0.250	–0.922	0.094	0.216	0.549
Factor 2 (mood)	0.414	0.086	0.605	–0.161	–1.295
Factor 3 (preparation convenience)	0.519	–0.084	–0.290	–0.215	–0.017
Factor 4 (purchase convenience)	0.400	–0.723	0.201	0.629	–0.496
Factor 5 (sensory appeal)	0.797	0.430	–1.017	–0.131	–0.264
Factor 6 (price)	0.366	0.039	0.489	–1.457	0.337
Factor 7 (weight control)	0.286	–0.156	–0.089	–0.142	0.072
Factor 8 (familiarity and ethical concern)	0.366	0.039	0.489	–1.457	0.337

Source: Reprinted from Milošević et al. (2012, p. 211). With permission from Elsevier

Table 4.2 Mean scores by cluster for variables used for cluster validation and profiling

	Cluster							
	1	2	3	4	5			
	Food enthusiast	Unconcerned	Price/distress	Purchase conven.	Health	Mean		
	$n=648$	$n=626$	$n=590$	$n=485$	$n=464$	$n=2{,}814$	*F*-test	Sig.
Consumption								
Fruit consumption[a]	37.9	34.3	33.3	36.4	37.3	35.8	4.6	***
Traditional dishes[a]	21.5	21.4	17.1	22.2	18.0	20.1	7.4	***
Health claim foods[a]	16.1	13.0	13.0	12.8	13.6	13.8	4.0	**
Knowledge								
Of organic food[b]	2.99	2.73	2.69	2.69	2.86	2.80	9.2	***
Of health claim foods[b]	3.08	2.70	2.85	2.86	2.90	2.88	12.9	***
Demographic/economic status								
Age	47.13	45.24	42.43	48.00	46.27	45.73	8.6	***
Income band[c]	8.30	8.30	7.72	9.41	8.09	8.35	8.0	***
Education band[d]	4.51	4.41	4.80	4.52	4.99	4.63	5.7	***
No. of household members	3.39	3.30	3.39	3.52	3.17	3.36	3.0	**
Respondent defined status								
Current state of health[e]	3.67	3.62	3.81	3.66	3.64	3.68	3.8	**
Econ position of household[e]	3.15	3.13	3.25	3.20	3.24	3.19	2.2	*
Categorical variables							X^2	
% Obese (based on BMI scores)	12.8	12.5	11.6	15.6	9.7	12.5	8.1	*

% Female	59.7	46.1	50.5	46.4	55.2	51.7	32.6	***
% With child(–ren) in household	38.6	33.5	40.8	34.6	37.5	37.1	8.1	*
% Pensioner	25.0	24.4	18.3	27.4	27.8	24.4	17.3	***

Source: Reprinted from Milošević et al. (2012, p. 212). With permission from Elsevier

***Significant at 1% level; **significant at 5% level; *significant at 10% level

[a]Measured in terms of portions/consumption experiences per month. For fruit a portion is approximately 80–100 g, for example, an apple, two spoons of fruit salad or 1 glass freshly squeezed juice

[b]Measured on a scale ranking from 1 = I am not informed at all to 5 = I am fully informed

[c]Measured in bands where a higher number equals a higher income band

[d]Measured in bands where a higher number equals a higher level of educational achievement

[e]Measured on a scale ranking from 1 = very bad to 5 = very good

Table 4.3 Cluster membership by country (%)

	Cluster membership				
	1	2	3	4	5
	Food enthusiasts	Unconcerned	Price/distress	Purchase convenience	Health oriented
	$n=648$	$n=626$	$n=590$	$n=485$	$n=464$
Country membership within clusters (%)					
BiH	30.2	18.8	14.1	18.2	18.6
Croatia	28.4	16.9	19.6	22.2	12.9
FRYOM	23.4	29.0	16.3	19.6	11.6
Montenegro	16.6	26.2	21.5	19.0	16.6
Serbia	25.7	20.0	29.4	15.5	9.4
Slovenia	13.5	22.7	24.8	9.8	29.1
Cluster total:	23.1	22.2	21.0	17.2	16.5

Source: Reprinted from Milošević et al. (2012, p. 212). With permission from Elsevier

current health status and economic position of the household are also closely in line with the respective means for the overall sample. This cluster reports highest level of fruit consumption and products with health claims. In this cluster, Slovenian and Montenegrin consumers are underrepresented.

4.4.2 Cluster 2: 'Unconcerned Food Consumers'

Compared to other groups, Cluster 2 places least emphasis on 'health and natural content'. The group also records below average mean scores for 'weight control', 'purchase convenience' and 'preparation convenience'. Overall their interest in food is weak. This group has the lowest proportion of females and percentage of households with children. The group has below average knowledge of organic foods and foods with health claims. Their economic status (income band) is close to the sample average. The group records below average consumption of fruit and products with health claims. Respondents from the Former Yugoslav Republic of Macedonia (FRYOM) and Montenegro are overrepresented in Cluster 2 whilst Croatia is underrepresented.

4.4.3 Cluster 3: 'Price-Oriented and Distressed Consumers'

Cluster 3, relative to the other groups, places greatest emphasis on 'price'. It records below average mean scores for 'weight control' and the lowest rating for 'sensory appeal'. This cluster places the highest relative emphasis on 'mood' (i.e. importance of *food to cope with stress*, *life*, *cheer me up* and *makes me feel good*, etc.). This cluster has the lowest average incomes. Average household size and composition are close to the means for the sample overall. Given their economic status, distressed consumers may turn to food for such comfort. Purchase convenience is perceived as relatively important, suggesting individuals that work long hours for little pay. This cluster reports the lowest average level of consumption of fruit and below average consumption of products with health claims. Their knowledge of organic and health claim foods is below the sample's average. Cluster 3 has above average membership of Serbian households.

4.4.4 Cluster 4: 'Purchase Convenience'

Compared to other groups, Cluster 4 places the greatest emphasis on 'purchase convenience' (*food that is easily available* and *can be bought in shops close to where I live or work*). This cluster records the lowest mean score for 'price' and registers the highest mean incomes. The cluster records below average scores for 'weight control' and 'familiarity and ethical concerns'. This cluster registers the highest level of consumption of traditional dishes, but the lowest consumption of products with health claims. Fruit consumption is in line with the overall sample average. This cluster exhibits the highest incidence of obesity (15.6%, measured by the body mass index (BMI)). Croatian consumers are most numerous in Cluster 4 (purchase convenience).

4.4.5 Cluster 5: 'Health-Oriented Consumers'

Compared to other groups, Cluster 5 gives greater relative importance to 'health and natural content'. The cluster records below average importance given to 'mood' and 'sensory appeal'. The low relative importance of 'mood' is striking. 'Price' is relatively important, and it appears this cluster will trade off hedonic pleasure for value and health benefits. This group has above average consumption of fruit. Their knowledge of organic food and health claim foods is, overall, above the mean for the sample as a whole. Self-reported health status is close to the average for the sample as a whole. Females are disproportionally represented in this cluster, and it has the highest percentage of pensioners. This cluster reports the lowest incidence of obesity (9.7%). Cluster 5 has a strong bias to Slovenia (29.1% of all members of the cluster), with comparatively few Serbs (comparable figure of 9.4%).

The profiling of clusters allows for the identification of groups that from a public health perspective raise concern. For the Western Balkans, three clusters raise concern. For each cluster, the communication messages should be distinctive, and these are described briefly below.

Cluster 2 (unconcerned consumers) displays relatively little interest in food and is biased to males. Food knowledge and fruit consumption are below average. Whilst there is a need to promote more balanced diets, this group may be fairly unresponsive to public health campaigns.

The group which presents the most concern is Cluster 3 (price-oriented and distressed consumers). This cluster places relatively more importance on the role of food for lifting their mood in the context of relative low average incomes. For these consumers, food may act as a comfort and stress reliever. Their level of fruit consumption is low. For these consumers, public health campaigns should stress the financial and health benefits from reducing fatty food intake.

Cluster 4 (purchase convenience) records the highest mean incomes and places the highest relative weight on availability and speed of shopping. This cluster displays the highest incidence of obesity. Rising incomes in the region may, as in other regions of the world, trigger an increase in demand for convenience foods and the prevalence of obesity. For the 'purchase-convenience'-oriented consumers, who report higher incomes but with the greatest incidence of obesity, the promotion of ready-to-eat light meals (e.g. fresh fruit or salads) would be appropriate (Buckley and Cowan 2007).

Having presented a broad overview of consumer segments, further work on improving the behaviour of at risk groups is warranted. This would look more closely at the efficacy of intervention strategies in the region, focusing on those clusters identified in this research that pose most concern.

4.5 Discussion of the Method: Utility and Encountered Difficulties

Whilst widely applied, cluster analysis presents a number of operational challenges. A critical issue is deciding on the number of clusters, for which there is no definitive rule. This means that the analysis depends heavily on the researcher's judgement. Second, formal procedures for assessing the validity and reliability of clustering solutions are complex and lack universal acceptance (Malhotra and Birks 2007). Malhotra and Birks (2007) recommend several procedures to check the adequacy of a clustering solution:

(a) Repeat the analysis with different distance measures, comparing the results to assess the stability of solutions.
(b) Repeat the analysis with different clustering methods, again comparing the results to assess the stability of outcomes.
(c) Split the data into two halves, undertaking the analysis separately on each half, then compare cluster centroids across the two subsamples.
(d) The order of cases may affect outcomes for nonhierarchical clustering. Repeat the analysis with cases entering in different orders until the solution stabilises.

These procedures were followed for the Focus Balkans dataset. The same number of clusters, with similar profiles and seed points, were apparent from the split-half approach.

4.6 Conclusions and Perspectives

CA is a useful technique to establish data groups and can underpin market segmentation studies. It involves a selection of method (hierarchical versus nonhierarchical or combination of both), interpretation to understand the character of each cluster and validation and profiling to check the stability of each cluster.

This chapter presents an illustration of the approach, namely, food motivations in the Western Balkans. Clusters are identified on the basis of responses to the FCQ, which provided a clear theoretical basis for the selection of variables. Five groups of consumers are discerned with the clusters characterised by significant differences in food consumption patterns as well as demographic and socioeconomic characteristics. Each cluster (food enthusiasts, unconcerned, price/distress, purchase convenience, health oriented) has a distinguishing set of motivations which are reflected in significant variations in food choice. Some of these clusters share characteristics identified in studies for other regions; for instance, Honkanen and Frewer (2009) identified a cluster of Russian consumers, with little interest in food, biased toward young males with low levels of educational attainment and income, which has much in common with Cluster 2 (unconcerned) for the Western Balkan Countries (WBC).

The analysis thus provides the basis for identifying distinctive groups of food consumers in the WBCs. Those groups with concerning patterns of food choice can be identified and appropriate public health messages and campaigns implemented. This recognises that the effectiveness of public health campaigns depends on tailoring the message to distinctive groups of consumers (Crossley 2001).

Recognising the weaknesses of CA, suggestions for further research can be identified. Repeating the study would allow for an assessment of the stability of clusters over time and gauging the degree to which particular groups are growing or reducing in size. For example, with economic development one may find that the size of the 'purchase convenience' cluster increases, whilst the 'price/distress' group shrinks. Further work could focus also on developing appropriate public health communications. This could involve focus group research, concentrating on relevant clusters and looking at reactions to different message appeals, exploring the reasons for such responses. This would be a logical next step in formulating communication messages and selecting media for tailored public health campaigns.

Lessons Learnt and Utility for Marketing Management

CA is a tool that can help managers segment a market and more accurately profile consumer needs and behaviour. The analysis for the WBC identifies five distinct groups of food consumers. These clusters transcend national boundaries and rather than thinking about, for example, the 'average' Montenegrin or Serbian consumer, it is preferable to consider the characteristics

(continued)

(continued)

of specific clusters and plan strategies that meet their needs. For instance, what are the needs and behaviour of 'purchase convenience' consumers? The CA reveals the diversity of food market segments in the WBC, so that strategies aimed at the 'average' or 'typical' WBC are flawed, as they are unlikely to meet the needs of any specific groups. CA helps marketing managers better understand the needs of specific segments, a vital step before deciding which segments to target and then how firms should position themselves in relation to chosen segments.

References

Askegaard S, Madsen TK (1995) Homogeneity and heterogeneousness in European food cultures: an exploratory analysis. In: European marketing academy conference proceedings. Gergy-Pontiose, France, 16–19 May 1995, pp 25–47

Buckley M, Cowan C (2007) Consumer attitudes towards convenience foods. In: Frewer L, Van Trijp H (eds) Understanding consumers of food products. Woodhead Publishing Limited, Cambridge, pp 200–218

Crossley ML (2001) Rethinking psychological approaches towards health promotion. Psychol Health 16:161–177

Everitt B, Landau S, Leese M, Stahl D (2011) Cluster analysis, 5th edn. Wiley, Hoboken

Geeroms N, Verbeke W, Van Kenhove P (2008) Health advertising to promote fruit and vegetable intake: application of health-related motive segmentation. Food Qual Prefer 19:481–497

Heinemann RJB, Behrens JH, Lanfer-Marquez UM (2006) A study on the acceptability and consumer attitude towards parboiled rice. Int J Food Sci Technol 41:627–634

Honkanen P, Frewer L (2009) Russian consumers' motives for food choice. Appetite 52:363–371

Kennedy J, Jackson V, Cowan C, Blair I, McDowell D, Bolton D (2005) Consumer food safety knowledge: segmentation of Irish home food preparers based on food safety knowledge and practice. Br Food J 107(7):441–452

Mai LW, Ness MR (1997) Consumers' perceptions of mail-order speciality foods. J Mark Manag 13(7):705–724

Malhotra NK, Birks DF (2007) Marketing research: an applied approach, 3rd European edn. Pearson, Harlow

Mesías FJ, Escribano M, Rodríguez de Ledesma A, Pulido F (2003) Market segmentation of cheese consumers: an approach using consumer's attitudes, purchase behaviour and sociodemographic variables. Int J Dairy Technol 56(3):149–155

Milligan GW, Cooper M (1985) An examination of procedures for determining the number of clusters in a data set. Psychometrika 50:159–179

Milošević J, Žeželj I, Gorton M, Barjolle D (2012) Understanding the motives for food choice in Western Balkan countries. Appetite 58(1):205–214

Murphy M, Cowan C, Henchion M, O'Reilly S (2000) Irish consumer preferences for honey: a conjoint approach. Br Food J 102(8):585–597

Punj G, Stewart DW (1983) Cluster analysis in marketing research: review and suggestions for application. J Mark Res 20(2):134–148

Romesburg HC (1984) Cluster analysis for researchers. Lifetime Learning Publications, Belmont

Saba A, Messina F (2003) Attitudes towards organic foods and risk/benefit perception associated with pesticides. Food Qual Prefer 14(8):637–645

Saunders JA (1980) Cluster analysis for market segmentation. Eur J Mark 14(7):422–435

Schnettler B, Vidal R, Silva R, Vallejos L, Sepulveda N (2009) Consumer willingness to pay for beef meat in a developing country: the effect of information regarding country of origin, price and animal handling prior to slaughter. Food Qual Prefer 20:156–165

Chapter 5
Exploring an Emerging Market through Focus Groups and Expert Interviews: Health Claim Products in the Western Balkans

Žaklina Stojanović, Galjina Ognjanov, Igor Spiroski, and Martine Zaouche-Laniau

Abstract The market of products with health claims is rapidly growing in the Western Balkan countries. This study has explored how the consumers value health claim products and what image they have of this product category. The following research questions were addressed: (1) Do consumers in Western Balkan countries associate health claim products with specific health benefits? (2) How do consumers perceive health claim products? Does the form of communication influence their choice of food? (3) What is the actual contribution of companies to this market's development? (4) How is a health claim product consumer perceived by different stakeholders? The purchase motives identified in this study are ranging from the belief in innovation and science over the favouring of specific ingredients such as probiotic bacteria to the appealing effect of, for instance, fat-burning characteristics.

Ž. Stojanović, Ph.D. (✉)
Department of Economic Policy and Development, Faculty of Economics, University of Belgrade, Kamenička 6, 11000 Belgrade, Serbia
e-mail: zaklina@ekof.bg.ac.rs

G. Ognjanov, Ph.D.
Faculty of Economics, University of Belgrade, Kamenička 6, 11000 Belgrade, Serbia
e-mail: galja@ekof.bg.ac.rs

I. Spiroski, M.D., MPH
Department of Physiology and Monitoring of Nutrition, Institute of Public Health, 50. Divizija 6, 1000 Skopje, Republic of Macedonia
e-mail: i.spiroski@iph.mk

M. Zaouche-Laniau
Sté GEM (Etudes et Stratégies pour l'Agroalimentaire), 58 A rue du Dessous des Berges, 75013 Paris, France
e-mail: mlaniau@gem.fr

D. Barjolle et al. (eds.), *Food Consumer Science: Theories, Methods and Application to the Western Balkans*, DOI 10.1007/978-94-007-5946-6_5,

Different scenarios are presented based on the results obtained through focus group discussions and expert interviews. This chapter explains in detail how these participative market research methods can be applied to a certain product category in the specific context of the emerging markets in the Western Balkans.

Keywords Focus groups • Health claim products • Western Balkans • In-depth interview • Qualitative methods • Healthy lifestyle

5.1 Introduction

The Western Balkan countries' (WBC) health claim products market is rapidly growing. Milk products with health claims were introduced in WBC more than 10 years ago. Development is based on the lessons learnt from the Western European economies. From the economic point of view, health claim products are often used as a differentiation tool in the development of food value chains. However, more responsive food systems are rather consumer than producer driven. Therefore, understanding the factors that affect health claim products choice (both of buying/consuming and producing/selling) is of crucial importance. The overall goal of this research is to identify the image and values that are given to the health food products in the Western Balkans. The analysis essentially refers to the market's current state and perspectives. Based on results obtained, different scenarios are presented. As the scope of possible future outcomes is observable, the scenario analysis gives more insights into alternative future developments.

Nutrition, as a lifestyle factor, has large impacts on human health. However, the significant increase in the cost of health care and social welfare is affected by rising rates of diet-related diseases. Consequently, government policies put more focus on health promotion and preventive measures against illnesses (Burrows et al. 1995; Petersen and Lupton 1996; Strauss and Dunkan 1998). Even now, most of the studies about foods with health claims in WBC can be found in the medicine and life sciences (Koch and Pokorn 1999; Šobajić 2002; Miletić et al. 2008). These studies cover the relation between nutrition and health as the core aspect of the research. However, health claim products have a huge potential to influence health outcomes, thus having important policy ramifications not only for public health policy and healthy lifestyle promotion but also for the food sector development. The studies of food consumers as well as of the food chain stakeholders' behaviour therefore seem to be particularly relevant for the emerging WBC health claim products market.

Finally, the countries included in this analysis (Bosnia and Herzegovina – BiH, Croatia, Former Yugoslav Republic of Macedonia – FYROM, Montenegro, Serbia, and Slovenia) significantly differ in the level of health claim product market development. The European Union (EU) Regulation of nutrition and health claims was adopted only in Slovenia and Croatia. However, national regulation in other countries is still not fully brought into agreement with EU legislation. The harmonisation will certainly change the situation.

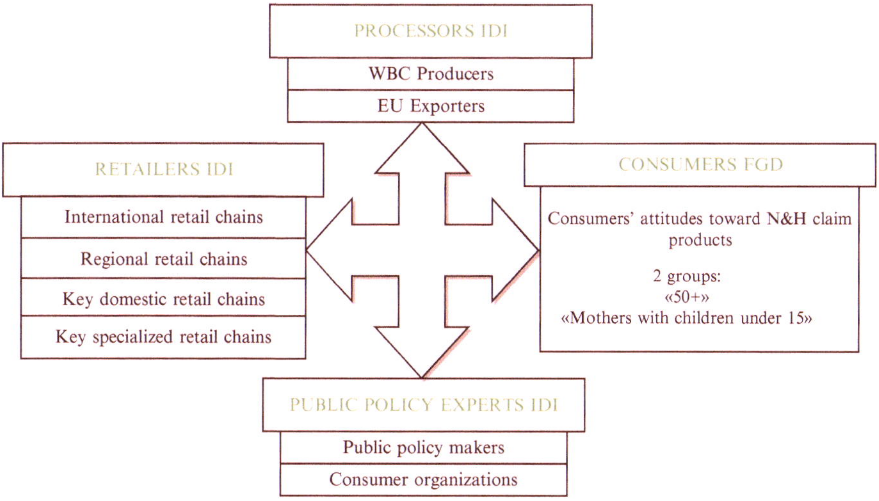

Fig. 5.1 Applied qualitative methods analysing the market stakeholders of health claim products in WBC. *IDI* in-depth interviews, *FGD* focus group discussions, *N&H claim* nutrition and health claim, *EU* European Union, *WBC* Western Balkan countries (*Source*: The Focus-Balkans research)

5.2 Method

The study of health claim products in the WBC is exploratory and consequently qualitative methods were used. These methods are designed to be the primary choice in the case of insufficient knowledge about the research topic (Hudson 2007). The specific methods were applied after having been adapted to each group of stakeholders (Fig. 5.1).

Focus group discussion (FGD) is the appropriate method whenever the group norms are observed, and the primary goal of the research is the discovery of richness of varieties in the population (Mack et al. 2005). It is the reason why FGD is the most appropriate method to capture consumers' perception of health claim products.

In contrast, *in-depth interview* (IDI) focuses on individual experiences usually addressed to highly sensitive topics. The studied market still remains a niche market with a small number of processors (although generally leaders of their market) involved in this domain in each country. Generally, health claim products have a small market share compared to the whole family of products (e.g. probiotic yoghurt compared to total yoghurt category). It was thus necessary to deepen the different topics with processors, retailers, and experts in the field.

The research study using qualitative techniques and methods presented in this chapter exclusively aims to explore both the consumers' and the economic sector stakeholders' experiences regarding the development of the WBC health claim products market.

5.2.1 *Origin and Purpose*[1]

IDI is a method based on the application of the psychoanalytical theory in marketing. Ernest Dichter pioneered this form of research (Solomon et al. 2010), promoting the idea that consumer behaviour is shaped by unconscious motives. Interviews have to be used to deepen insights into each person's motivations and have to be interpreted according to Freudian psychoanalytic concepts.

Understanding the context of the stakeholders' behaviour is the main goal in interviewing. In a business environment, in-depth interviewing is mainly lead by the interest in a deeper understanding of experiences of interviewed stakeholders. Researchers usually observe behaviours under the following basic assumption: the meaning interviewees make on their business experiences affects the ways that carry out that experience (Seidman 1991).

Widely used in many sciences, among which psychology and sociology, IDI became a qualitative research technique aiming at deeply exploring people's thoughts, feelings, and perspectives on particular economic issues in the business area. In marketing research, IDI may be used for two main purposes: (1) to analyse deeply consumers' attitudes and behaviour and (2) to analyse experts' knowledge on a specific market. Usually carried out on small samples, these interviews primarily address issues of creative nature and motivational factors that are drivers of behaviour (Lai Ngun et al. 2009; Bruggen et al. 2011).

In the case of IDI in the industrial environment (e.g. in market research), the sampling is different from a quantitative survey based on a quota method. IDI refers to stratified sampling (Mack et al. 2005). The population is divided into few subgroups or strata (e.g. large companies, medium ones, and small ones). Then a sampling rate is decided on for each stratum. Practically, strong efforts should be put into gathering all the leaders of the studied market, while a specific proportion can be carried out with smaller or less present companies/stakeholders in the market segment. In our case, IDI with the food chain stakeholders appears to be an essential tool both to analyse the processors'/retailers'/policymakers' experiences and to evaluate the main market tendencies.

5.2.2 *Material and Data Processing*

FGD served as a tool for acquiring information regarding consumers' images and values related to health food products in the Western Balkans. The specific topics selected for FGD were (1) concepts of healthy lifestyle, healthy diet, and their relation; (2) food purchase and food consumption; (3) purchase process: information that makes a difference for buyers; (4) perception and understanding of nutrition and health claims; (5) motives and barriers for purchase and consumption of products

[1] An extensive presentation of the FGD method is provided in Chap. 6. Therefore this chapter refers only to IDI.

with health claims; (6) perception of a typical consumer of products with health claims; and (7) future expectations regarding products with health claims.

Based on the literature review, two groups of people were chosen for FGD: elderly (50+) and mothers who have children younger than 15 years old. The elderly usually have health problems that can be prevented or mitigated by a specific diet and healthy nutritional habits (Moorman and Matulich 1993; Bech-Larsen and Grunert 2003). However, young children are an especially vulnerable group, and studies conducted recently show that mothers usually pay more attention to question what their children eat (so-called role of a gate keeper) than other population segments (Wansink 2005). Therefore, groups that seemed particularly product related or health concerned were selected. A total of 98 participants took part in FGD from December 2009 to January 2010 in all six observed countries. Discussions were conducted in six WBC capital cities.

The WBC health claim products market survey was based on the experts' IDI (processors, retailers, public authorities, and consumer organisations). The specific questionnaires were adapted to each type of stakeholders. Processors, retailers, and experts were asked about (1) overall attitude towards consumers' demand, (2) main criteria and restraints for consumers' choice, (3) perceived consumers' characteristics, and (4) insights into studied categories (general presentation and estimation). Public policy experts' point of view was concentrated on (1) nutritional concerns and consumer awareness, (2) variety and nutritional quality of the offered product, (3) regulations and compatibility with EU legislation, (4) public programmes of communication to promote better food habits, and (5) priorities and the main governmental objectives in the area of research. All stakeholders were asked about the trends and perspectives of the health claim products market. IDI were conducted with the top managers (brand or sale/marketing directors of companies) as well as with the policymakers from ministries of agriculture/health and consumer organisations in the region.

The list of the companies was obtained through shop survey conducted in September 2009. The list consisted of 41 identified firms. Totally 29 WBC companies took part in interviewing during November 2009 to January 2010, and 7 interviews were conducted with the most important exporters of EU health claim products. Additionally, 26 food-retailing companies, both the leaders and followers, were interviewed. The public policy stakeholders' experiences were obtained through 18 interviews (three per country in the region). These sums up to a total of 80 interviews with food sector stakeholders. The diversified structure of respondents enables the high reliability of the study.

5.2.3 Utility and Encountered Difficulties

Our study exclusively looks for results that explain the recent evolution of the WBC health claim products market both from the consumers' and food sector stakeholders' perspective. The market structure, consumer demand, and trends are observed.

However, the implementation of IDI in an industrial environment encountered some problems. First, the competition among the agro-food industries and the large retailers hinders the disclosure of data and figures. It was important to guarantee that all individual data from companies would be kept strictly confidential and would be published with great caution in order to prevent the recognition of companies' data and respondents. Second, appointment with adequate respondents was very difficult. It is important to stick to adequate respondents because the method is based on the high level of knowledge of the interviewees, although these important individuals are neither easily accessible nor always willing to answer any survey. Third, when implementing the IDI, there is a must to take into account that the companies' representatives usually tend to overestimate their sales, market share, and more generally the performance of their company. Therefore, it is necessary to double-check the data. To achieve this verification, several methods can be combined: previous data gathering (surveys or studies), cross-checking of data gathered from different stakeholders during interviewing, etc. Finally, a generalisation based on the IDI results is usually not allowed due to small samples and non-random sampling. However, when several companies in a given sector including leaders of the market have been met, it is reasonable to assume that a good perception of the market and its tendencies has been captured.

5.3 Context of the Study and Research Questions

Epidemiological transition in terms of shifting the public health attention from communicable to chronic non-communicable diseases (CNCD) is characteristic for the second half of the twentieth and the beginning of the twenty-first century. The prevalence of CNCD is rapidly increasing which makes them one of the top public health priorities. Cardiovascular diseases (CVD) are the world's number one reason for death. It is estimated that 18 million people died from CVD in 2008, representing 40% of the total deaths in the world (World Health Organisation – WHO 2008). CVD, along with cancer, which is the second main cause for death globally, is closely related to nutrition (WHO 2011). Moreover, four of the ten leading causes of death in the United States are directly related to nutrition (Sizer and Whitney 2011). As for the situation in the WBC, the age-standardised death rate (SDR) from circulatory diseases for all ages per 100,000 inhabitants ranges from 231.77 in Slovenia to 531.80 in Macedonia. The values for other studied WBC are 386.08 for Croatia, 461.63 for Montenegro, and 506.61 for Serbia. The average of this rate among EU member states (member states before May 2004) is 180.84 (WHO 2011; Institute of Public Health – IPH 2010). Risk factors that may lead to CNCD are divided into two groups: the ones we cannot modify, control, and treat and the other ones that are modifiable. Preventable risk factors include smoking, high blood cholesterol, high blood pressure, physical inactivity, overweight, obesity, and diabetes mellitus. Other factors, which may also contribute to high CNCD prevalence, are individual responses to stress and excessive alcohol drinking (American Heart Association – AHA 2011).

Keeping in mind the epidemiological situation explained above, this study addresses the following research questions: (1) Do consumers in WBC associate health claim products with specific health benefits? (2) How do consumers perceive health claim products? Does the form of communication influence consumers' choice of food? (3) What is the actual contribution of companies (both processors and retailers) to the development of the health claim products market? (4) How is a health claim product consumer perceived by different stakeholders? The final output of this research illustrates the motives and behaviours towards health claim products in the Western Balkans from consumers', producers', retailers', and policymakers' points of view.

5.4 Results of FGD and IDI

The results presented hereafter are related to diet and health issues, motivations and barriers for the consumption and production of health claim products, perception and understanding of claims, and to the identification of the typical health claim product consumer as well as of expectations regarding the market development.

5.4.1 Healthy Diet and Healthy Lifestyle: How Important Is the Health Motive in WBC Food Consumption?

FGD shows that healthy food is among the first concepts associated with healthy lifestyle for both elderly people and mothers with children up to 15 years old in all WBC. Healthy food is generally connected to the intake of fruits and veggies as well as to cooked (i.e. warm meals) and home-prepared food and diet diversity. However, in Slovenia, 'physical activity' is especially important to elderly participants (primary association, before healthy food) who added mental activity as another important concept.

Elderly man, 63, Slovenia:

> After retiring, people could freeze on sofa.

'Cleanness' is mostly related to the environment and environmental issues (clean water, clean and fresh air, etc.) contributing to the production of healthy food (Montenegro, Macedonia, and Serbia).

Typical associations to the concept of a healthy lifestyle are[2]:

Healthy food: fruit, vegetables, cooked – homemade (warm meals), food diversity

Sport activities: walking, running, biking, hiking, swimming, dancing

(continued)

[2] *Source*: Consumer statement from FGD, the FOCUS-BALKANS research.

(continued)

Environment: clean air and earth, ecology, naturally grown food
Healthy life routines: regular sleeping, frugality, and moderation in all activities
Economic and social security: employment, financial safety
Interpersonal networks: friends, neighbours, communication, hobbies
Positive atmosphere: warmth, love, family, loyalty, support
Dissipation: absence of cigarettes, alcohol, drugs, any kind of distress

Economic and social security are frequent associations with healthy life in being the primary prerequisites for a life without stress. For all FGD participants, healthy lifestyle stands for harmony in family life, interpersonal networks with people who can support a person when he or she needs it, and economic and social security. Economic and social security are particularly stressed as an important precondition for a life without stress.

Elderly woman, 58, Macedonia:

> It is essential to have economic well-being in order to be healthy.

Parent, 36, Slovenia:

> It is important that you receive a check each month in order to have a healthy life.

In BiH, economic and social security was also related to political stability as yet another important contributor to a life without stress, i.e. a healthy life. Finally, in some countries the respondents related healthy lifestyles to certain more psychological or even spiritual values, such as optimism, voluntary work for others, unselfishness and giving rather than receiving in general (Serbia), religion, meaningful life, meditation, yoga (Slovenia), and positive atmosphere without political misunderstandings (Montenegro, BiH).

The study tries to find out consumers' perceptions on the link between food and health. Their associations covering different kinds of food, food quality and food safety issues, sports and sportsmen, nature, beautiful women, happy families and babies may be viewed as typical (Fig. 5.2).

The experts – producers, retailers, policymakers, and consumer groups' representatives – mainly agree that price is the main problem that consumers actually face. This problem is most pronounced in the least developed WBC – Macedonia and BiH.

Producer, BiH:

> Health benefit is very important to the consumers. More important than we usually think, but they are restricted with the price and cannot always afford these products.

Producer, Macedonia:

> Consumers are interested in healthy products only if their price is equal with the regular ones.

Fig 5.2 The collage technique (examples) (*Source*: The FGD conducted in the Focus-Balkans research)

Additionally, experts from Croatia and Serbia pointed out that price might be an important obstacle for the lower income households from less-developed rural areas. High prices were one of the problems accentuated in Slovenia, but this has particularly been obvious since the outbreak of the economic crisis. Besides price, processors and retailers pointed out taste as the second most important criteria for consumers' choice of food.

Producer, Serbia:

> Consumers are traditionally looking for good taste even if it is not good for their health.

Consumer perception that food which contributes to a good health is not tasty was especially emphasised by the public health policy officials. This is often the reason why consumers refrain from buying healthy products.

The facts mentioned above confirm that health benefit might not be a dominant criterion for consumers' choice of food in WBC. However, the health officials emphasised that healthy diet became the leading factor for consumers' choice of food in specific situations – e.g. in the case of children (newborns) or family members who have got some health problems.

Public Health Policy Official, Croatia:

> All parents want to feed their children with the highest quality food as much as it is possible. But when adult people buy food for themselves high quality is not so important, price is more important.

Nevertheless, health benefit was perceived as a factor of growing importance.

Retailer, Macedonia:

> Consumers are quite aware of the health benefits of some food. They are buying products with health and nutrition claims even they don't have any health problems.

Retailer, Croatia:

> Consumer's demand for products with nutrition and health claims will grow due to health issues – such as increasing cardiovascular diseases.

5.4.2 *The Basic Market Drivers: Motivations and Barriers for Consumption*

According to FGD, consumers' motives for choosing products with health claims can be divided into several categories: (1) innovation, scientific background; (2) containing specific ingredient (familiar health claim, e.g. 'probiotics improve digestion'); (3) prior personal experience with the product; (4) word of mouth on product effects; and (5) appealing effect (e.g. 'burns fat').

On the other hand, there are strong barriers to adopt them: (1) doubt in added ingredients: vitamins, minerals, Q10, and other enzymes that are untraceable; (2) some additives are unnatural and unnecessary (e.g. vitamins in yoghurt, calcium in fruit juice); (3) health claim is not well communicated to consumers (too long, too technical, too complicated); and (4) products are perceived as targeting certain consumers, and consequently, some consumers do not view themselves as potential consumers. There are two main causes to these barriers: first, the majority of WBC consumers is still uneducated when it comes to this type of products and expresses resistance towards interpreting complicated scientific claims; second, there is doubt and refusal to accept a novelty (health claims that are longer present on the market are more taken to be trustworthy, while the new ones are more often rejected).

Furthermore, the consumers' likelihood analysis refers to food choice. Different groups evoke different motives for consumption: dairy products are consumed on an everyday basis, and, therefore, consumers are more aware of their features (e.g. low fat, probiotics). Elderly respondents in all WBC claim to be making their choices based on their health and diet benefits implying that they are paying more attention to nutrition and health claims. However, the respondents mostly prefer the products they are familiar with (well-established 'traditional' brands) and/or which taste better. Traditional brands are the first choice particularly in the group of mothers who claim not to be ready to experiment with what they give to their children. Thus, a specific role of the branded products of well-known companies in novelty acceptance in WBC is confirmed.

The companies considered products with health claims to be important for the creation of their image. Improving a company's image by selling healthy food and answering to the consumers' needs were the motives most often evoked spontaneously both by processors and retailers.

Retailer, Serbia:

> We are selling safety products with health claims. It has influence on our consumers, their trust in our retail chain.

Producer, Serbia:

> Expanding the target group of consumers, sale increase as well as improving the image of company, indirectly influence the consumption of other product categories.

The companies use health claim products as a marketing positioning tool and do not expect immediate high profitability in the WBC market. Some categories of health claim products still generate modest sales volumes compared to the conventional market share. As far as the economic sector stakeholders' point of view is concerned, main barriers for the development of the health claim products market are consumer related. According to producers and retailers, unhealthy eating habits followed by low perceived awareness result in a low demand. Simply said, consumers include these products insufficiently in their diets as they are neither aware of them nor of their role in an overall healthy diet.

From a public policy perspective, the health policymakers confirm these findings as well. According to their assessment of the current situation, the consumers are not sufficiently educated about nutritional food contents and a healthy diet. They are not able to understand food declarations and therefore are not able to choose high-quality products according to labelling information.

The general assessment of the interviewed experts is that nutrition and health claims are attracting the consumers to buy a product by highlighting its benefits. However, they might mislead, mostly non-educated, consumers if they are insufficiently precise and not verified scientifically. This is most likely to be the case in the Western Balkans, where the health claims market is still not fully regulated in all regions. However, due to a harmonisation with the EU regulation, consumers trust the scientifically substantiated claims increasingly.

5.4.3 The Perception of a Typical WBC Health Claim Product Consumer: The Reality or a Vision?

From the consumers' point of view, a typical consumer of health claim products is seen as a young woman, usually working and with a higher income. She is well educated, friendly, and an extrovert. As for physical appearance, she is in a good shape, fit, and cares a lot about her appearance. She is usually described as married with children, but she has also been seen as a single. She buys products in large supermarkets on a weekly basis. She follows trends and seeks for novelty in every domain (Fig. 5.3).

The processors' and retailers' perception of WBC health claim product consumers is quite identical. Consumers are generally female, belonging to the age group of 15–40 years old, with higher or middle income, secondary or higher education, with or without health problems, and are living in urban areas (Fig. 5.4). Consumers of health claim products are health oriented – their everyday practice is connected to healthy lifestyles, following modern trends and fashion in food consumption (they want to stay fit) and active life (sportsmen, businessmen), or they are seen as mothers who are expected to provide healthy food to their families.

A typical health claim product consumer image in WBC seems to be heavily influenced by television commercial stereotype. Additionally, elderly people usually see a typical health claim product consumer as a young, active female in her 20s. Mothers sometimes see an elderly woman with health issues but trying to stay fit

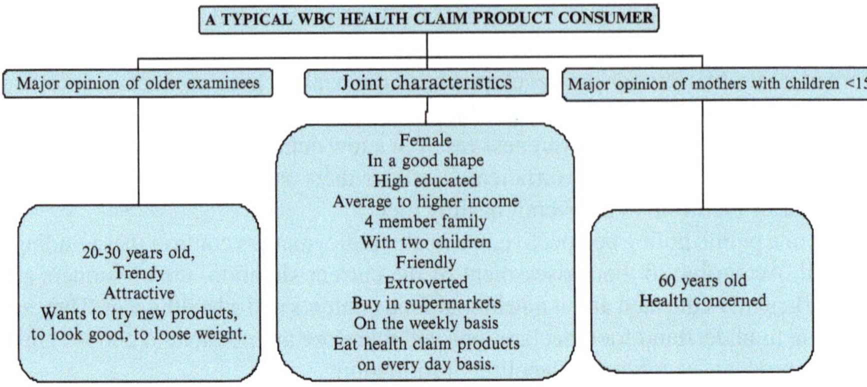

Fig. 5.3 FGD results: a typical WBC health claim product consumer seen from the point of view of a consumer (*Source*: The FOCUS-BALKANS research)

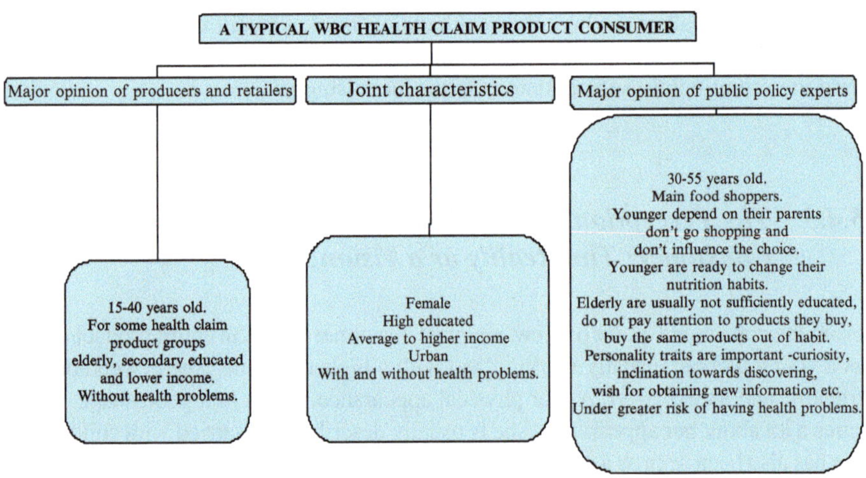

Fig. 5.4 A typical WBC health claim product consumer seen from the point of view of producers, retailers, and policymakers (*Source*: The Focus-Balkans research)

and taking care of herself as a typical consumer. It is important to note that each age group perceives functional food to be typical for the other, which might also indicate a lack of actual identification with these products.

Furthermore, important differences are identified between the business sector and policymakers' perceptions (Fig. 5.4). The policymakers insist more strongly on consumers with health problems and under a greater risk of having health problems as a typical health claim product consumer than the business sector stakeholders who allocate them in a much broader context.

5.5 Conclusions

In Montenegro, Macedonia, BiH, and Serbia, food with health claims is perceived to be an upcoming trend in the food market, while it is an actual trend in Croatia and Slovenia. Consumers do not have spontaneous expectations regarding the future development of health claim products. When asked, they say they expect (1) more consumer awareness (health impact of different ingredients to be more commonly known), (2) greater availability (in regular supermarkets, not specialised stores), and (3) a more diversified offer. However, Kapsak et al. (2011) found that even those consumers who are more and more informed and knowledgeable about health claim products are still more likely mentioning products that may have health benefits rather than the specific components of these products.

The highly regulated health claims product market could be an efficient dissemination tool for the health and risk-diseases information to the general public. Additionally, the impact of health professionals has been identified as critical to consumers' success in healthy eating. The food industry typically produces health claim products by incorporating functional elements into processed foods or supplements. However, strong scepticism is expressed by our experts because they think that those elements are present in amounts too low to result in appreciable health benefits. For that reason, e.g. for the purpose of CVD prevention, diets rich in antioxidants and phytochemicals rather than high-dose supplementation are still recommended. Moreover, the benefits of health protective, whole grain foods are assessed to be greater than those of functional food products.

It is important to stress that consumers and health professionals need to inform and educate themselves in order to keep pace with the ongoing scientific advancements in this area and to be able to assist the consumers when they include these products in their diets. Health professionals might have a significant influence on specific population segments such as mothers of newborns, children, elderly, and those who experience diet-related health problems. Active public policy in the field of research can promote the concept of healthy lifestyle and healthy eating in the region. Last but not least, investigations whether the promotion of health claim products could contribute to a change in the overall diet towards healthier food choices should lead to a general improvement of the food chain competitiveness. Thus, further research should address quantitative analyses of both heavy and light consumers' profiles in the WBC. Explanation of attitudes, believes, and motives for health claim products consumption is equally important. Identification of the main predictors of healthier food choices can reshape diet patterns in the WBC and promote health claim products as an efficient tool for improving public health. Research based on qualitative methods contributed to the *Food Consumer Science* development by identifying the main driving forces on the market of the study area.

Lessons for Food Marketers: Main Insights into the Market Development of Products with Health Claims in WBC

Some studied categories of products with health claims were introduced in the WBC more than 10 years ago. However, the number of products with health claims has been constantly increasing since the year 2000. Consumer demand, and thus the global market, for health claim products is expected to grow in the future for several reasons. Consumers' awareness regarding the importance of food will increase as the stakeholders (producers, retailers, consumer organisations, and public policymakers) have great interest to improve the nutritional education of the population. Due to the harmonisation with the EU regulation and the control that European Food Safety Authority (EFSA) and WBC Food Safety Agencies will carry out on the market, the faith of consumers in the scientifically substantiated claims is going to increase. New products are also expected to enter the markets: risk disease reduction claims will be authorised in all WBC, and this will strongly influence the markets (e.g. authorised cholesterol lowerers in margarines).

Based on the qualitative data obtained by FGD and IDI, two scenarios for the market development of WBC products with health claims can be identified. According to the first scenario, which is pessimistic, the health claim products market pass through a fashion life cycle. The target population consists of young, educated, high income, and urban women as the main consumers. However, in more developed markets (Slovenian and Croatian) the target population is now extended to mothers who are seen as the households' gatekeepers. In the second, more optimistic, scenario, elderly population with higher and middle income along with innovators and early adopters (young active women and mothers) also become interesting consumer groups for products with health claims. The athletes and businessmen are seen as potential consumers in the future.

The market is obviously going through a multidimensional transition – from the younger to the elderly population, from the higher income class to the middle class layers, from women to men. The food sector stakeholders play a crucial role in adopting and implementing new technologies which may contribute to an increasing availability of these products both in terms of diversity and price. Simultaneously, consumer associations and policy makers take leading-strings in defining the institutional framework that is going to facilitate further developments of the WBC market.

References

AHA (2011) Risk factors and coronary heart disease. Retrieved 14 Aug 2011, from AHA: http://www.heart.org/HEARTORG/Conditions/HeartAttack/UnderstandYourRiskofHeartAttack/Understand-Your-Risk-of-Heart-Attack_UCM_002040_Article.jsp

Bech-Larsen T, Grunert KG (2003) The perceived healthiness of functional foods: a conjoint study of Danish, Finnish and American consumers' perception of functional foods. Appetite 40(1): 9–14

Bruggen A, Ranjani K, Sedatole KL (2011) Drivers and consequences of short-term production decisions: evidence from the auto industry. Contemp Account Res/Rech Comptab Contemp 28(1):83–123

Burrows R, Nettleton S, Bunton R (1995) Sociology and health promotion. Health, risk and consumption under late modernism. In: Bunton R, Nettleton S, Burrows R (eds) The sociology of health promotion. Routledge, London, pp 1–9

Hudson D (2007) Agricultural markets and prices. Blackwell Publishing, Oxford

IPH (2010) Health and health protection of the population of the Republic of Macedonia. IPH, Skopje

Kapsak WR, Rahavi EB, Childs NM, White C (2011) Functional foods: consumer attitudes, perceptions, and behaviours in a growing market. J Am Diet Assoc 111(6):804–810

Koch V, Pokorn D (1999) Comparison of nutritional habits among various adult age groups in Slovenia. Nutr Res 19(8):1153–1164

Lai Ngun S, Kay R, Chew M (2009) Development of a retail life cycle: the case of Hong Kong's department store industry. Asia Pac Bus Rev 15(1):107–121

Mack N, Woodsong C, MacQueen K, Guest G, Namey E (2005) Qualitative research methods: a data collector's field guide. Family Health International, Research Triangle Park

Miletić I, Šobajić S, Đorđević B (2008) Funkcionalna hrana – uloga u unapređenju zdravlja. J Med Biochem 27(3):367–370

Moorman C, Matulich E (1993) A model of consumers' preventive health behaviors: the role of health motivation and health ability. J Consum Res 20(2):208–228

Petersen A, Lupton D (1996) The new public health: health and self in the age of risk. Sage Publications, London

Seidman IE (1991) Interviewing as qualitative research: a guide for researchers in education and the social sciences. Teachers College Press, New York

Sizer FS, Whitney E (2011) Nutrition: concepts and controversies, 12th edn. Wadsworth, Belmont

Šobajić S (2002) Funkcionalna hrana u prevenciji bolesti i terapiji. Arhiv za Farmaciju 52(3): 369–375

Solomon M, Bamossy G, Askegaard S, Hogg MK (2010) Consumer behaviour: a European perspective, 4th edn. Financial Time Press/Prentice Hall, Harlow

Strauss J, Dunkan T (1998) Health, nutrition and economic development. J Econ Lit 36:766–817

Wansink B (2005) Marketing nutrition: soy, functional foods, biotechnology, and obesity. University of Illinois Press, Urbana/Chicago

WHO (2008) World health report 2008. WHO, Geneva

WHO (2011) Cancer linked with poor nutrition. Retrieved 14 Aug 2011, from WHO: http://www.euro.who.int/en/what-we-do/health-topics/diseases-and-conditions/cancer/news/news/2012/3/cancer-linked-with-poor-nutrition

Chapter 6
Assessing Fruit Perception Using Focus Groups

Siet J. Sijtsema, Karin L. Zimmermann, Miljan Cvetković, Cristina Mora, and Martine Zaouche-Laniau

Abstract Today, only a minority of the world's population consumes the recommended daily amount of fruits and vegetables. Even though fruit growing has a long tradition in the Western Balkan countries, fruit consumption should be encouraged as it is a basic element of a healthy diet. During the FOCUS Balkans project, fruit perception and consumption was studied with a qualitative and quantitative approach. This chapter describes the application of focus group discussions as a widely employable method in food consumer science.

Required preparation measures and guidelines for focus group discussions are explained in detail. Furthermore, practical aspects about the implementation of this method as well as advice on the selection of participants are presented on behalf of a concrete example. For instance, it seems that participants from the Western

S.J. Sijtsema, Ph.D. (✉)
Consumer and Behaviour Group, LEI, Part of Wageningen UR,
P.O. Box 35, 6700 AA Wageningen, The Netherlands
e-mail: siet.sijtsema@wur.nl

K.L. Zimmermann, B.Sc., R.M.
Consumer and Behaviour Group, LEI, Part of Wageningen UR,
P.O. Box 29703, 2502 LS The Hague, The Netherlands
e-mail: karin.zimmermann@wur.nl

M. Cvetković, Ph.D.
Faculty of Agriculture, University of Banja Luka,
Bulevar vojvode Petra Bojovića 1A, 78000 Banja Luka, Republika Srpska-BiH
e-mail: miljan.cvetkovic@agrofabl.org

C. Mora, Ph.D.
Department of Food Science, University of Parma,
Via Kennedy 6, 43039 Parma, Italy
e-mail: cristina.mora@unipr.it

M. Zaouche-Laniau
Sté GEM (Etudes et Stratégies pour l'Agroalimentaire),
58 A rue du Dessous des Berges, 75013 Paris, France
e-mail: mlaniau@gem.fr

D. Barjolle et al. (eds.), *Food Consumer Science: Theories, Methods and Application to the Western Balkans*, DOI 10.1007/978-94-007-5946-6_6,

Balkans generally prefer home-grown fruits or fruits bought on green markets, and fruit safety was an issue frequently raised. Focus group discussions hence are a valuable tool in exploring the consumption of a specific product, especially if combined with a quantitative survey for investigating differences between countries or between people of different age groups.

Keywords Application of focus groups • Fruit perception • Western Balkan consumers • Home-grown • Home-made • Safety

6.1 Introduction

As basic elements of the daily human diet, fruits and vegetables enable a healthy life and prevention from diseases. Only a minority of the world's population consumes the recommended average amount of fruits and vegetables of 400–500 g/day (World Health Organization (WHO) 2003). The consumption of fruits in Europe has been declining over the last years. Only a few Mediterranean countries, where availability of fruit is high, are currently meeting the public health recommendations.

To elaborate public health policies, more insight into the preferences of European consumers regarding fruits and derivative products is necessary. Especially the position of fruit consumption in transition countries such as the Western Balkan countries (WBC) is of interest, in order to develop well-planned interventions to encourage fruit consumption. There is a particular need to explore consumers' motivations and barriers as well as the background of consumption and purchase of fruits. To this end, exploration of as well as deeper insights into the perceptions of consumers is needed. During the FOCUS Balkans project, fruit perception and consumption was studied in a qualitative and quantitative approach. This chapter is about the application of focus group as a method used in the food consumer science, applied to two specific topics related to fruits: perception of motives for and barriers to different channels for getting or buying fruits and consumers' associations with national fruits. The aim was to better understand consumers' perceptions and expectations about fruits, to promote fruits and produce fruit innovations.

6.2 Method

6.2.1 Theoretical Background

In consumers' daily life, there are several variables influencing the choice for food, of which some are described in Chap. 1.

In a number of consumption models, the determinants of product choice are structured according to the three determinants: the individual, the situation and the product (e.g. Belk 1975; Bloch and Richins 1983; Dickson 1982; Zaichkowsky 1985). Similarly, these same determinants are used to structure variables related to food behaviour and perception (e.g. Gains 1996; Kahn 1981; Randall and Sanjur 1981; Rozin 2007; Sijtsema 2003).

Individual determinants of food perception could be divided into demographic variables, physiological factors, psychological factors and attitudes. Food-related determinants of food perception are decomposed into product characteristics (e.g. the colour of the food) and the production system (e.g. whether the food is organic or not). In addition, while most studies refer in the situational component to both physical surroundings and social surroundings (Belk 1975), Sijtsema (2003) made a further distinction between environment and context. Whereas the environment consists of the social and cultural environment (family and society characteristics), the context contains the consumption moment and place of consumption (Sijtsema 2003). According to Köster and Mojet (2007), situations are defined by the meaning attributed to the surroundings and the resulting expectations of the individual with regard to this situation. For example, hunger means that we want to eat something, but hunger when sitting alone in front of the TV reminds of other types of food than when at a dinner with friends at home or with a group of business relations at a restaurant. It is clear that people do not just eat products, but they eat what they like in different meaningful situations (Köster and Mojet 2007). In Sijtsema's (2003) model, no distinction is made between the objective aspects (the package is blue and it contains 200 g of rice) and the perceived aspects (the product is tasty). Rozin (2007) notes that ultimately the food product and environment are filtered through the person: that is, it is the *perceived* product and environment that influence choice.

6.2.2 *Choice of Method*

From a food consumer research perspective, the focus of this chapter is on the exploration of consumers' perception as the background of consumption and purchase of fruits. The perception of consumers is based on a stimulus, which is differently perceived according to the consumer characteristics, time and situation. Focus groups are a method often used to explore consumer perceptions in which it is possible to take those characteristics of consumers and contexts into consideration.

6.3 Origin and Purpose

Focus groups can be defined as 'a research technique that collects data through group interaction on a topic determined by the researcher' (Morgan 1997, p. 6). Historically, the focus group method emerged at the Bureau of the Applied Social

Research of Columbia University. Merton and Kendall (1946) pioneered the approach with their work about the effectiveness of propaganda efforts during the Second World War. They discovered that group discussions allow a broader range of response and elicitation of important and informative details that would not be mentioned with individual interviews.

The method has been developed since the 1990s, with a systematic description and ways of implementation such as how to conduct focus groups and analyse data (Krueger 1988; Morgan and Krueger 1998; Stewart and Shamdasani 1990; Templeton 1994; Vaughn et al. 1996).

6.4 Material and Data Processing

Focus groups can be implemented at many points of a research programme. Often, they are applied to prepare quantitative research. Focus groups are helpful to generate a hypothesis that can be validated with further quantitative approaches (Stewart and Shamdasani 1990). Besides focus groups' facilitation of quantitative research tools, they are also helpful to broaden and deepen insights of, for example, in-depth interviews. Focus groups are originally a specific tool for qualitative data collection, based on the dynamics of the group. Although the group process is one of its advantages, it might also result in an undesirable bias in the group. To avoid biases, one should be keen on the following issues. First, the moderator has an influence on the group. Second, as in all human groups, the phenomenon of dominance may occur and some participants may strongly influence the discussion. The following actions should be taken to avoid this bias: (a) development of a very precise guideline, (b) trained moderator, who is aware of and knows how to deal with group interactions and dominant participants, and (c) balance between individual and group tasks.

Focus groups bring rich and sensitive information about a topic of interest, which led to an increasingly more abundant use of them as a research method on its own. The risk of 'misuse', especially in the case of market research, is pointed out by Merton (1987, p. 557) 'as quick and easy claims for the validity of the research are not subjected to further quantitative test'. Results of focus groups are not meant to be representative and no generalisation of findings can be made on a specific population. Implementing multiple focus groups does not allow the generalisation of findings but could allow the researcher to gain additional information and also to reach a kind of saturation of information (information occurring repeatedly). Thus, one should be careful when applying focus groups as a method on its own since one could not reach representativeness.

One of the advantages of focus groups is their flexibility. Researchers adapt the procedure according to their own needs. Those needs should be carefully considered and translated to questions or tasks. The guideline of the focus group discussion needs careful preparation, strictly adapted to the research objectives. In order to allow the emergence of all ideas and opinions, different tasks and questions as well as different tools will be applied to stimulate the group in this process. Pictures, products and videos can be usefully presented to give a support to the group's representations and reactions. Also, indirect methods or projective techniques such as games or specific exercises can be used. Consumers generally appreciate such exercises and the discussion is more vivid and creative while researchers may explore more underlying representations.

6.5 Practical Aspects

6.5.1 Protocol for Focus Groups

There is not a single recipe to build the protocol of focus groups.

The guide of the discussion is generally designed in such a way that we can move from general points of view to very precise ones. Broadly outlined, the discussion can be divided into four or five parts:

The first part is a 'warming up' phase. It is important to create an open atmosphere at the beginning of the session. Therefore, it should be explained that there are no right and wrong answers. A round table with short presentations of participants and researchers is an opportunity to support consumers to feel confident. In this phase, exercises or questions on the general topic are important for starting to understand the subject.

A second phase may be devoted to the exploration and mapping of the issues related to the general topic of the following phases aiming at answering precise questions. In the area of food consumption, they can be oriented towards purchase habits, frequency of consumption, distribution channels for the studied product or towards families of products.

The last phase aims at deepening specific research questions. Participants are now very familiar with the topic. At this stage, it could be asked to clearly elicit their opinions, preferences and decisions regarding precise questions. At last, synthesis exercises may be proposed for example on the possible marketing arguments, opportunities or recommendations.

6.5.2 Preparation and Moderation

Focus groups need careful preparation long before the meeting itself (development of guidelines, recruitment of participants and renting a room). This also has consequences for the budget to be planned: hiring of the room, office equipment, catering and consumers' compensations (gift or money).

If possible, the moderator, not necessarily the same person as the researcher, should have a small pilot group of 3–4 participants, so he might test the questionnaire and estimate the time needed for the task realisation. The session cannot reasonably exceed 2 h. If more time is needed, a break might be necessary to foster the active involvement of the participants.

The moderator of the discussion plays an important role during the meeting. He or she has to be familiar with the guideline and must respect the timing for each question and exercise. The moderator should make participants feel free and comfortable and avoid any dominance phenomenon among them. The whole discussion is audio or video recorded. If needed, transcriptions have to be prepared in order to do the analyses and reporting. A complete synthesis of the 'raw material' given by the group has to be done. Analysis and interpretation are rather time-consuming, as it is crucial to take into consideration all ideas and opinions produced by the group. Conclusions are drawn question by question in order to reflect the diversity and heterogeneity of the data.

It is often useful to conduct at least two focus groups; in theory, one should do as many focus groups until no more new issues emerge. It is possible to conduct focus groups for two different age classes or locations (rural/urban) or any other different characteristics. Moreover, it is recommended to conduct in each case at least two similar focus groups to gain a 'saturation' of information. When cross-country differences are explored, the groups should be conducted with the same protocol in each country.

6.5.3 Participants of Focus Groups

Consumers may be recruited from the general population aiming at a well-balanced group regarding gender, age or education level. Focus groups can also be constituted with a specific population segment having generally no more than one characteristic that distinguishes them from others (not only socio-demographics) depending on the research project (e.g. heavy users of organic food).

The ideal number of participants in a group ranges from around six to ten persons. Organisers need to over-recruit (12 persons) in order to replace people who will cancel the appointment at the last minute. It should be avoided that participants know each other.

6.6 Research Approach and Objectives

Perception and consumption of fruits in everyday life have been studied by focus groups in the Focus Balkans project. Fruit was defined as fresh fruit (e.g. apple, banana, tangerine) and fresh pressed juice. Dried fruits (grape, apricots, figs, plums) and other processed fruits such as juice, jam and frozen fruits are considered as a category by itself. However, processed food that contains low contents of fruits, such as fruit yoghurts, biscuits with fruits or fruit ice creams, is not considered in this study. Also non-eatable fruits, nuts and pastes are not considered as fruit.

Fig. 6.1 The qualitative research approach and the position of focus groups

Central in this part of the study are consumers and their motivations and behaviours relating to fruits and fruit products in the Balkans. To this end, several activities were carried out to gain a qualitative understanding of fruit consumption in the WBC. At first, in-depth interviews with consumers as well as in-depth interviews with experts from companies and government were scheduled. Besides, focus groups were carried out to further explore some specific topics arisen in the in-depth interviews, of which some results will be presented in this chapter for being an illustration of a tangible application of focus groups (Fig. 6.1).

The objective of the focus groups was to explore the outcomes of the interviews with regard to (1) motives and barriers for fruit consumption and its availability at different channels, for example, home-grown fruit, fruit from the supermarket or from green markets, and (2) the perception of national fruits and fruits from other countries. A selection will be presented in this chapter.

Each WBC (Bosnia and Herzegovina (BiH), Croatia, Former Yugoslav Republic of Macedonia (FYROM), Montenegro, Serbia and Slovenia) took care of the data gathering in the respective country. For urban as well as for rural participants, two separate focus groups with all together 96 participants were conducted. The sample was balanced for gender and region, with an overrepresented group of students. A transcript was written in the local language of each group discussion. Based on this, the results were presented in an elaborated format and translated to English.

6.7 Results

Home-grown fruits have a tradition in the Balkan countries (Delic and Zimmermann 2010). This tradition might influence the perception of fruits (Sijtsema and Snoek 2010). In times of transition and urbanisation, it is interesting to find out how it influences the perception and purchase of fruits in Western Balkan countries. Therefore, this study explored in-depth consumer perception of different distribution channels of regional, national and international fruits.

6.7.1 Perception of Home-Grown Fruits, Green Markets and Supermarkets

Generally, participants prefer the green market for buying fruits. Fruit safety is a frequent issue raised by the participants: buying fruits on green markets is a way to be sure about the safety of the fruits (which means in consumers' opinion no treatment with pesticides), and that the fruits are locally produced, sold by a person they know. After the green markets, the second most liked place of procurement is the own home in the rural area, while supermarket shopping is preponderant in urban areas. Table 6.1 shows the broad variety of all kinds of motives and barriers mentioned during the focus groups in the different countries. The motives and barriers of buying fruits at different channels are grouped for home-grown fruits, green markets and supermarket in order to show the diversity of issues mentioned. The home-grown fruits and green markets have been grouped for representing consumers' habits in buying fruits.

The discussion shows that participants from rural areas prefer home-grown fruits, and that their most preferred place to buy fruits is the green market. This seems also to be preferred by urban participants but they also buy fruits in the supermarket since the distance to the supermarket is shorter and availability of fruits is better. Some participants never will go only for fruits to the supermarket. It seems that urban participants are more familiar with supermarkets than rural participants. This seems to be connected with the small number of supermarkets in rural areas in some WBC. Some people have a habit with regard to a specific sales channel while others buy at different channels. It seems that elderly people prefer green markets more often. Younger people think supermarkets are more convenient. For some participants, price is an issue.

It seems that participants trust individual sellers more than big supermarket chains. There is a general scepticism of fruit safety. Fruits are believed to be overtreated with pesticides. In the rural area, they prefer their own production, which is by their experience not sprayed. Generally, consumers have problems with credibility attributes.

This description shows that participants mention several issues related to where they buy their fruits, for example, supermarket or green market. If one should know the importance of issues compared to others, either as differences between countries or target groups, additional quantitative research is needed.

6.7.2 National and International Fruits and Their Perception

The perception of national and international fruits is studied in order to better understand the general perception of fruits, the position of fruits as a part of traditions and the buying characteristics of fruits.

Many different fruits are perceived to be 'national' fruits in Western Balkan countries. Apple is often mentioned as a national fruit for all WBC. Tangerines are

Table 6.1 Motives and barriers of buying fruits at different channels, for example, green markets or supermarkets

	Motives home-grown/green market	Barriers home-grown/green market	Motives supermarket	Barriers supermarket
Montenegro rural	Less treated Affordable price Local – to support production and purchase conditions Better quality Healthier (we know how it is grown) Cleaner Fresher I know the people who sell it	Fruit is not treated and gets spoiled quickly Not sufficient amounts sold quickly Not many producers of home-grown products Availability Not well-organised purchase	Lack of time Easier access Availability whole year Compensates for fruits which are not available in the season or fruit which is not grown in own country Longer storability	Worse appearance Forced growing, not tasty fruit Price, expensive Less quality Lack of confidence in quality and origin Treatment changed taste
Montenegro urban	Fresh fruit in the morning/ season Know people, confidence More tasty and fresh	Far away from house, only on weekend No time to go Not enough fruits Not well stocked Working hours at green market overlap with own working hours	Convenient supply More fruits available Longer opening hours than green markets Supermarkets are everywhere Lack of time Working hours Nearest shop Various fruits Cheaper than green market	Fruit not fresh enough Fruit seems unnatural Less tasty Artificial Price
Macedonia rural	More natural atmosphere Bigger choice Always same seller	No big offer Time	Specific types of fruits	Not a big choice More expensive
Macedonia urban	No pesticides used Fruit at green market is closer to home-grown	I do not want a crowded place where someone is convincing others that he has the best products	In neighbourhood More freedom, no suggestions from sellers Tropical fruits available	Less fresh Not sure where fruit comes from and if it is fresh or not Doesn't like to shop in supermarket
Bosnia rural	Petrol costs			

(continued)

Table 6.1 (continued)

	Motives home-grown/green market	Barriers home-grown/green market	Motives supermarket	Barriers supermarket
Bosnia urban	Good assortment When you buy more you get discount	High temperature in summer, fruits get rotten You cannot choose the fruit yourself Not many green markets in the city Awkward when it rains No frozen fruit	Availability Get fruit along with regular weekly purchases You can choose the fruit yourself Facilities to store the fruit are better Visual arrangement of fruits in shop	
Serbia rural	Direct contact with fruit and grower Modern markets in Belgrade offer same fruit products Not chemically treated Try the taste of fruit before buying	Weather conditions Storage conditions Market too crowded Unpleasant smells everywhere. Our farmers are still not well educated; I am still not 100% sure about quality	Everything in one place Even in winter exotic fruit More attractive Pay with credit card	Bananas are not kept in a proper way Unknown country of origin, transportation time conditions of storage unknown
Serbia urban	When I buy green market I help some villager	Sanitary inspection should do a better job in green market		
Croatia urban			Great offer in supermarket Buys fruit here in winter much more often while many fruit markets do not work Poor quality Unripe fruit	Less healthy because you do not know where it is from Fruit is not fresh it is in a freezer Do not like that other people touch fruit Imported fruit does not have a good taste and looks artificial
Croatia rural				Long in stock before it is sold

mentioned as the most popular national fruit for Croatia and plums for Serbia (cf. Table 6.2). Participants choose their national fruit for diverse reasons like growing the fruit in their own garden or seeing it a lot on the green market. Also traditions influence the perception, for example, processing of plums into brandy in this region. Besides, participants mention frequently that national fruits are of higher quality than imported ones and contain fewer pesticides. Domestic fruit production is considered to be healthier, to taste better and the domestic fruits easier to preserve than other fruits. When discussing this, participants mention that they think that the Balkan region has good conditions for fruit growing, which is related to traditional recipes and traditional varieties of fruits.

Although this exploration shows differences among the countries, confirmation is needed through a quantitative approach: in fact, the focus groups allow identifying extensively all issues, which consumers associate with national fruits.

6.8 Discussion

6.8.1 Application of Focus Groups

When interpreting the results, we have to be aware that a qualitative research approach was followed. This means that consumers' perceptions, ideas and experiences of and attitudes towards the research topic, in this case fruits, are discussed in the focus groups. This is of value when there are nearly no studies available. After exploring the topic qualitatively, a quantitative research is needed to test the findings for reasons of representativeness. Thus, each of the six countries organised two focus groups with altogether 96 Western Balkan people, balanced for gender and region. Comparison of the countries seems interesting but only should be done by means of a quantitative study to reach a valid comparison. Nevertheless, interesting differences were explored and need to be investigated further in a quantitative study in order to test the expectations.

Especially in a project like Focus Balkans where six countries each with their own language had to be coordinated in English, some extra barriers had to be taken. For example, the translation of the guide to local languages and of the transcript to English had to be tackled on time.

The application of focus groups was not a common method for some partners of the WBC. Nevertheless, there were good reasons for its use. Therefore, additional trainings about moderating focus groups were organised.

6.8.2 Richness of Data

For each topic discussed, a lot of issues were mentioned by participants. Although not all are similarly relevant, each should be taken seriously as representing the

Table 6.2 National fruits per country

Country	'National' fruit	Issues participants mentioned and discussed about national and international fruits
Montenegro	Apples Grapes Plums Less often: pears, figs	Many types of fruits can be grown in Montenegro Tradition of making processed fruit products and alcoholic beverages
Macedonia	Apples Grapes Less often: peaches, plums	Good conditions (a lot of sun) for growing fruits in Macedonia If possible, people would prefer to buy only Macedonian fruits Macedonian fruits are the best because they are less treated with chemicals 'Fruits are something we should be proud of'.
Bosnia	Apples Plums Pears Less often: peaches, strawberries, cherries	Traditional species as well as new species Contradictions about quality: one respondent says 'soil is poisoned' while another says 'far less pesticides in Bosnian fruits than in imported ones' Traditional fruits are produced in a safe and controlled manner, therefore it is a better quality than fruit produced on big farms or imported fruit 'Musmula' is a national fruit 'Plum was the national fruit, but now it isn't anymore'
Slovenia	Apples Pears Cherries Plums Less often: grapes	Slovenian fruits are associated with tasty, healthy, juicy, fresh, good appearance, and grown in a soil, which is not as polluted as in other countries Some traditional fruits are not known anymore, for example, 'nesplje' Domestically grown fruits are preferred because of freshness, better quality, less time spent in storage There is also some influence of national feelings
Serbia	Plums Apples Pears Less often: cherries, raspberries	Respondents had the opinion that Serbian fruits are unsurpassed. Climate is extremely appropriate for fruit production; it is proven by high yields of fruits National fruits are healthy and they are easily available and cheap and delicious products can be made of it It could be preserved all over the winter time
Croatia	Tangerines Apples Plums Less often: strawberries, grapes, figs	Although participants do not pay attention to the origin of the fruits they buy, they announce preferring to buy Croatian rather than imported fruits

voice of a consumer. Even though most of the issues might be in line with prior expectations, additional insight into the background of the issues and into experiences of participants in their daily life as well as into their motives and barriers for consuming fruits were valuable to better understand consumer food choice behaviour.

6.8.3 Reflection on Results

The insights from these focus groups are valuable for formulating hypotheses, which can be tested in quantitative research. Several interesting topics emerged, for example, whether there are differences between rural and urban consumers as far as their buying behaviour with regard to fruits is concerned and whether there are differences in the perception of safety of the products.

Fruits seem to be part of the traditions in the Western Balkans which is shown by the high diversity of varieties and species and by abundant recipes. Participants prefer the fruits from their own country or region, but it is not known whether they really take this into consideration when they buy fruits. Due to the transition in the Western Balkan region, there are several developments going on which might influence the perception of consumers. The position of domestic production of fruits and of fruit purchases on green markets is changing since supermarkets have appeared. Before, fruits were bought primarily on the green markets where sellers were at the same time producers with rather small family farms. At that time, existing markets did not or only rarely organise the sale of fruits. Those small producers were not ready for the change of selling to supermarkets since they could neither offer the required quantity of goods nor fulfil the norms of supermarkets demand. Nowadays, big supermarkets in WBC offer mainly fruits from export (European Union, EU) or from domestic plantations that are able to produce enough goods according to the requested standards. These fruits cannot be considered as 'home-grown'. The basic reasons for the connection between home-grown fruits and green markets mainly lay in the tradition of fruit buying but also in the role of supermarkets.

Agro-food industry in WBC is more and more oriented to a production that satisfies the requests of big supermarkets and to some additional requests for the export of goods to the EU countries and Russia. In WBC, consumers' habits regarding fruit purchases will change and be more oriented towards supermarkets. There are several reasons for this: less and less people are living in rural areas but are moving to urban areas; buying in supermarkets is becoming a dominant way of fruit buying; and agro-industry is investing in fruit production for supermarkets and has reliable partners for selling its products.

The entrance of foreign retailers might thus influence the perception of fruits and possibly also fruit consumption in a society with changing lifestyles and ongoing urbanisation. Public policy might be needed to avoid a further decrease in fruit consumption in WBC.

6.9 Concluding Remarks

Focus groups are a promising way to explore the perception of fruits, although complementing quantitative studies are needed to investigate differences between countries or between people of different age groups.

Focus groups open many options for further research within the field of consumer perception since they proved to be a suitable approach for the exploration of issues and new phenomena related to the process of perceiving fruits.

Focus groups are a valuable tool in exploring fruit consumption, especially if they are combined with other methods such as a quantitative survey for better understanding the consumers in order to create innovations for fruits and the fruit sector as well as to tailor interventions from public authorities.

The future of the agriculture and the food industry in the Balkans depends on the producers' ability to understand global and regional market trends. It would be important to take consumers' needs and perceptions into consideration because these are not in line with the developments going on in the market at the moment. To avoid a further decrease of fruit consumption, authorities might be needed to accompany people in this process.

References

Belk RW (1975) Situational variables and consumer behavior. J Consum Res 2:157–164

Bloch PH, Richins ML (1983) A theoretical model for the study of product importance perceptions. J Mark 47:69–81

Delic A, Zimmermann K (2010) Deliverable 5.1 Report on the production and consumption of fruits and fruits products in the Balkans. Project no 212579, FOCUS BALKANS. Retrieved 25 July 2012, from Focus-Balkans Web site: http://www.focus-balkans.org/?p=69

Dickson PR (1982) Person-situation: segmentation's missing link. J Mark 46:56–64

Gains N (ed) (1996) The repertory grid approach. Measurement of food preferences. Blackie Academic & Professional, London

Kahn MA (1981) Evaluation of food selection patterns and preferences. CRC Crit Rev Food Sci Nutr 15(2):129–153

Köster EP, Mojet J (2007) Theories of food choice development. In: Frewer L, van Trijp H (eds) Understanding consumers of food products. Woodhead Publishing Limited, Cambridge, pp 93–124

Krueger RA (1988) Focus groups: a practical guide for applied research. Sage, London

Merton RK (1987) The focused interview and focus groups: continuities and discontinuities. Public Opin Q 51:550–558

Merton RK, Kendall PL (1946) The focused interview. Am J Sociol 51:541–557

Morgan DL (1997) Focus groups as qualitative research, vol 16, 2nd edn, Qualitative research methods series. Sage, London

Morgan DL, Krueger RA (1998) The focus group kit. Sage, Thousand Oaks

Randall E, Sanjur D (1981) Food preferences – their conceptualisation and relationship to consumption. Ecol Food Nutr 11:151–161

Rozin P (2007) Food choice: an introduction. In: Frewer L, van Trijp H (eds) Understanding consumers of food products. Woodhead Publishing Limited, Cambridge, pp 3–29

Sijtsema S (2003) Your health!?: transforming health perception into food product characteristics in consumer-oriented product design. Dissertation, Wageningen UR, Wageningen

Sijtsema S, Snoek H (eds) (2010) Deliverable 5.2 study report on consumer motivations and behaviours for fruits and fruit products in the Balkans. Scientific report. Project no 212579, FOCUS BALKANS

Stewart DW, Shamdasani PN (1990) Focus groups: theory and practice. Sage, London

Templeton JF (1994) The focus group: a strategic guide to organizing, conducting, and analyzing the focus group interview, revised edn. Probus Publishing Co, Chicago

Vaughn S, Schumm JS, Sinagub J (1996) Focus group interviews in education and psychology. Sage, Thousand Oaks
World Health Organization (WHO) (2003) Diet, nutrition and the prevention of chronic diseases. Report of a Joint WHO/FAO Expert consultation. WHO technical report series 916, Geneva
Zaichkowsky JL (1985) Measuring the involvement construct. J Consum Res 12:341–352

Chapter 7
Harnessing Expert Opinion: Trends and Challenges on the Balkan Organic Market, Based on a Delphi Approach

Natasa Renko, Ružica Butigan, Sanda Renko, Ante Vuletić, Burkhard Schaer, and Martine Zaouche-Laniau

Abstract The market for organic products in Western Balkan countries has only recently started to develop. Today, it is still a small niche market. This chapter presents an analysis of the organic food market in the mentioned countries based on the knowledge and opinions of experts. The main objectives of this study were to identify and aggregate the opinions of experts regarding the current state of the Western Balkans organic food market and to forecast the future trends and challenges as well as to suggest improvements for this market.

Data were gathered using the Delphi method. This approach allows a group of experts to deal with a complex problem. Basically, the Delphi method is a series of

N. Renko, Ph.D. (✉)
Department of Marketing, Faculty of Economics and Business, University of Zagreb,
Trg J. F. Kennedyja 6, 10 000 Zagreb, Croatia
e-mail: nrenko@efzg.hr

R. Butigan, M.B.A. • A. Vuletić, M.A.
Faculty of Economics and Business, University of Zagreb,
Trg J. F. Kennedyja 6, 10 000 Zagreb, Croatia
e-mail: rbutigan@efzg.hr; avuletic@efzg.hr

S. Renko, Ph.D
Department of Trade, Faculty of Economics andj Business, University of Zagreb,
Trg J. F. Kennedyja 6, 10 000 Zagreb, Croatia
e-mail: srenko@efzg.hr

B. Schaer, Ph.D.
ECOZEPT, Oberer Graben 22, 85354 Freising, Germany
e-mail: schaer@ecozept.com

M. Zaouche-Laniau
Sté GEM (Etudes et Stratégies pour l'Agroalimentaire),
58 A rue du Dessous des Berges, 75013 Paris, France
e-mail: mlaniau@gem.fr

D. Barjolle et al. (eds.), *Food Consumer Science: Theories, Methods and Application to the Western Balkans*, DOI 10.1007/978-94-007-5946-6_7,

structured questionnaires, commonly referred to as rounds. Answers of one round are used as feedbacks in subsequent rounds. The approach taken in this study is described in detail with a special emphasis on the selection of experts and items for the questionnaires.

The three Delphi rounds showed that the organic market in the Western Balkan countries is expected to increase in the future and that education and information of consumers hereby are key success factors. While most experts agreed on these main findings, their opinions varied on other issues raised in the questionnaires.

Keywords Organic food sector • Western Balkan countries • Delphi method • Expert panel • Consumers' motivation

7.1 Introduction

During the last decade, from the year 2000 onwards, the organic movement around the world experienced new dynamics (Žakovska-Biemans 2011). The global market for organic products has increased by 235% since 1999, with Europe and North America holding the leading position in that growth. Countries with the largest market of organic products are the USA, Germany, the United Kingdom and France (Sahota 2010). The mentioned trend has been spread to different extents in Western Balkan countries (WBC). This chapter presents an analysis of the organic food market in the mentioned countries based on expert opinion gathered according to the Delphi approach.

Delphi has been proven to be a useful method for eliciting expert opinions within the food domain (Soon et al. 2012; Wentholt et al. 2009, 2010). The method allows a group of individuals, as a whole, to deal with a complex problem (Linstone and Turoff 2002). It could also be used to seek out information which may generate a consensus from the respondent group or to explore underlying assumptions or information leading to different judgements (Hsu and Sandford 2007; Linstone and Turoff 2002). The Delphi technique is essentially a series of structured questionnaires, commonly referred to as rounds (Henson 1997), where answers are used as feedback in subsequent sessions or rounds. By providing feedbacks, Delphi creates interactivity and a dialogue without the pressure of conforming to the group's decisions or to influential respondents (i.e. the group's position being overly swayed by dogmatic or high-powered individuals). Nonprobability sampling (i.e. purposive sampling or criterion sampling) is generally accepted as appropriate in Delphi studies as the opinion of experts is sought (Powell 2003). From a practical perspective, Delphi reduces costs and allows access to many individuals across diverse locations and a wide distribution of panellists (Soon et al. 2012; Wentholt et al. 2010). The aim of applying the Delphi technique in this study was twofold. The first aim was to identify and aggregate the opinions of experts regarding the current state of the organic food market in Western Balkan countries. Second, the experts also serve to forecast the future trends and challenges as well as to suggest improvements.

7.2 Organic Food Sector in WBC

In the last decade, the interest in organic agriculture has grown both within the scientific community and among the general public. Organic agriculture is an agricultural production system that tries to fully utilise farming potential to satisfy all the social and economic needs while preserving the natural ecosystem and ensuring environmental protection (Colom-Gorgues 2009). The most important characteristics of organic products are the control system of organic farming and certification and the labelling of the products. Literature review shows that an increased consumer awareness of food safety issues and environmental concerns have contributed to the growth in organic farming over the last decades and years (Colom-Gorgues 2009). According to the latest data (Cerjak et al. 2010), consumers in certain WBC consider health effects of organic food and care for the environment among the most important motives for buying organic food products. Moreover, the organic production is related to sustainable rural development. Sustainable development must encompass food production alongside the conservation of limited resources and protection of the natural environment so that the needs of people living today can be met without compromising the ability of future generations to meet their own needs (Colom-Gorgues 2009).

The production of organic products in the WBC has started to develop recently. The market for organic products is still a small niche market. The degree of market penetration of organic products is still limited and at a very low level but with expectations of growth. Table 7.1 presents the main differences between the developed European countries and the WBC.

The main difference between the selected countries lies in purchasing power. Average gross domestic product (GDP) per capita (purchasing power parity, PPP) of WBC is below US $15,000, except for Slovenia and Croatia. This directly affects the consumption of organic food. While in developed European countries, for example, Denmark, consumers are spending more than Euro (€)140 annually on organic food, the average amount spent in the WBC is around €6 with Slovenia being the leader. Significant differences can be observed in organic production as well. While the share of organic land in developed European countries exceeds 3% (Denmark 5.4%), the WBC average is around 1% with Slovenia being an exception. Its share of 5.9% corresponds with European trends in organic production (Willer and Kilcher 2011).

Bearing in mind all differences, the objectives of this study are twofold—to better understand (a) the current situation and (b) the future and potential evolution of organic in WBC through an experts' survey.

Table 7.1 Main differences in the organic sector between developed EU countries and WBC

Statistical indicators	Developed European countries	WBC
GDP per capita-PPP (average in USD)	30,388	14,000
Share of organic land	3%	From <1 to 6% (Slovenia) Around 1% in average[a]
Consumption (annual amount spent on organic food per capita in USD)	From 15 to 150 Around 80 in average[a]	From 1,2 (Bosnia) to 20 (Slovenia)[b]

Sources:
[a]Specialised organic retail report Europe 2008. ORA, Ecozept, Biovista (projections 2010) et Agence Bio (2011)
[b]Schaer (2012)

7.3 Method

7.3.1 Delphi-Based Approach

7.3.1.1 Origin and Purpose

We used a forecasting methodology in order to achieve our goals. For our purposes, we selected the Delphi method that is based on a series of written questionnaires with feedback and re-voting. The method structures and facilitates group communication that focuses on a complex problem with the purpose of achieving group consensus about the future direction over a series of iterations. The sample used in the Delphi method is based on a panel of carefully selected experts that represent a wide spectrum of opinions on the topic being examined. The statements are usually anonymous (Loo 2002).

Since its development by the Rand Corporation to improve (military) technology forecasting in the 1960s, Delphi studies have been applied extensively. The approach involves successive questionnaires to an expert panel, using feedbacks to refine an informed perspective on complex or uncertain issues. Epistemologically, Delphi studies are not merely deductive but also disclosive (Jones 1989) and allow fragmentary perspectives to coalesce into a larger collective understanding. The best definition of the approach is found in the seminal work of Linstone and Turoff (1975, p. 3):

> Delphi may be characterised as a method for structuring a group communication process in such a way that the process is effective in allowing a group of individuals, as a whole, to deal with a complex problem.

A review of this methodology and critiques on Delphi studies are provided by Mullen (2000). A small number of studies have been carried out to forecast food market development, but as Critcher and Gladstone (1998) note, its use in applied social science is not widespread. The objective of most Delphi applications is the reliable and creative exploration of ideas or the production of suitable information for decision making. The Delphi method is based on a structured process for collecting and distilling knowledge from a group of experts by means of a series of questionnaires interspersed with controlled opinion feedback (Adler and Ziglio 1996). It is an exercise in group communication among a panel of geographically dispersed experts. The technique allows experts to deal systematically with a complex problem or task. The essence of the technique is fairly straightforward: it comprises a series of questionnaires sent either by mail or via computerised systems to a preselected group of experts. It can be applied in other ways as well: by telephone, by face-to-face interviews or by mixtures of different survey methods. The number of rounds and the size of the group can vary. The main intention of the Delphi method is to overcome the disadvantages of conventional committee action. According to Fowles (1978), anonymity, controlled feedback and statistical response characterise Delphi studies. A statistical response is not always necessary, qualitative results are possible as well. The group interaction in Delphi is anonymous in the sense that the originator of comments, forecasts and the like is not identified.

In the original Delphi process, the key elements were (1) structuring of information flow, (2) feedback to the participants and (3) anonymity for the participants. These characteristics may obviously offer distinct advantages over the conventional face-to-face conference as a communication tool. The interactions among panel members are controlled by a panel director or monitor who filters out material not related to the purpose of the group (Martino 1983). The common problems of group dynamics are thus completely bypassed.

The method is also criticised, for instance, that Delphi studies are unscientific (Sackman 1975) or that accuracy is not given (Armstrong 1978). Furthermore, it is useful for answering one, specific single-dimension question. There is less support for its use to determine complex forecasts concerning multiple factors. Such complex model building is more appropriate for quantitative models with Delphi results serving as inputs (Gatewood and Gatewood 1983). This point is supported by Gordon and Hayward (1968) who claim that the Delphi method, based on the collation of expert judgement, suffers from the possibility that reactions between forecasted items may not be fully considered. The need for integrating the cross impact matrix method of forecasting into the Delphi method is pointed out by many researchers (Gordon and Hayward 1968; Gatewood and Gatewood 1983; Adler and Ziglio 1996). An improvement in forecasting reliability over the Delphi method was thought to be attainable by taking into consideration the possibility that the occurrence of one event may cause an increase or decrease in the probability of occurrence of other events included in the survey (Helmer 1977). Therefore, cross impact analysis has been developed as an extension of Delphi techniques.

We adapted the Delphi method to our specific case. There are three participant roles: the design group that creates the questionnaires for the expert group; the expert group that answers the questionnaires; and the analysis group that makes the consensus of the expert group (Rodriguez-Diaz 2000). The Delphi method procedure envisages conducting more than two rounds of written tests about selected items in the organic food industry (Rodriguez-Diaz 2000). Having done so, we generated and rated options and projections by our expert group that was composed of 6–10 people per country. We then retained consensus results in confidence.

7.3.1.2 Sampling and Selection of Experts

Sampling and selection was done in accordance with the recommendations and experience found in the literature (e.g. Soon et al. 2012). Expert panels were invited to take part in the Delphi study for identifying and selecting the most relevant problems and challenges in the organic food sector in WBC. Here, the panellists were not selected randomly, so representativeness is not assured. The selection of experts for the Delphi study was made through the following:

- Personal contacts of the authors and the research supervisory committee
- Participants of international organic food conferences
- Experts co-nominated by other experts (Scapolo and Miles 2006)

7.3.2 Data Collection and Processing

A total of 45 experts from the organic food sector were contacted and invited to participate in the Delphi survey. Experts were defined according to two criteria: (1) currently teaching a university-level food science, agriculture or aquaculture programme or working in the agriculture or organic sector, and (2) experienced in the organic food sector. The invitation contained a cover letter with a short description of the study and the questionnaire. Even though it is more advantageous to conduct a face-to-face interview in the first round to increase the response rates, it could not be done for all countries in this study due to the limited financial resources and time. Three rounds of questions and answers were deemed to be optimal (compare Soon et al. 2012).

Table 7.2 shows the number of participants who took part in the three rounds of the Delphi study and the respective gender balance. The response rates were high as

Table 7.2 Sample characteristics of Delphi rounds I, II and III

	No. of experts			Gender	
Country	Round I	Round II	Round III	F	M
Bosnia and Herzegovina	6	6	6	3	3
Croatia	8	8	8	6	2
Macedonia	4	3	3	2	1
Montenegro	3	5	5	2	3
Serbia	8	8	8	5	3
Slovenia	10	14	14	6	8
Total	39	44	44	24	20

it is typical for Delphi studies (Soon et al. 2012; Grundy and Ghazi 2009; Stärk et al. 2002; Wentholt et al. 2010).

7.3.3 Round 1

The first round of the Delphi study consisted of a highly structured open questionnaire based upon extensive review of the literature (Hsu and Sandford 2007). In a classical Delphi, round I begins with an open-ended set of questions that allow panellists to generate ideas (Soon et al. 2012). There is a potential for bias due to the limited available responses (Keeney et al. 2001), but participants were given the option to list additional problems and challenges of the organic food sector in WBC. Round I's questions were divided into six main sections: (1) influences in the organic food market in the last 10 years, (2) the current state of the organic market, (3) expectation for the next 10 years, (4) motives in consumer behaviour, (5) impact of organic food initiatives and (6) organic food distribution channels.

The survey for the first round started at the end of April 2010 and was finished by the beginning of May 2010. Afterwards, the report was prepared and sent to the same group of experts in order to get first feedbacks. Once completed, experts were required to return the questionnaire to the author for statistical aggregation and review.

7.3.4 Round II

The feedback from round I was aggregated and the central tendency and dispersion of scores were indicated to the participants in the subsequent round (Soon et al. 2012). It was also indicated to participants where their scores stood in relation to the overall picture. This enabled them to revise previous scores. In round II, experts were required to review the feedbacks of round I and revise their scores if necessary.

At this stage a second questionnaire was applied. It consisted of 229 questions and was split into 6 different parts of common thematic contents titled as: (1) Government and Policy Impact, (2) Production, (3) Market, (4) Trends, (5) Supply Chain and (6) Consumer Behaviour. All the answers from the first round were compiled to identify core topics for the second round. The survey for the second round started at the end of February 2011 and was finished by the beginning of March 2011.

7.3.5 Round III

The feedback and revised scoring from round II was aggregated and reviewed. In round III, the experts were required to answer an additional set of questions from the second round that had met a low level of consensus. Round III questions were sent to the experts who had responded in round II.

There is a need to emphasise that the basic criterion for drawing conclusions in this part of the research study was the level of consensus on offered items in the questionnaires. The main task was to identify questions with a predominant support (or reached a level of consensus above 50%) from the experts. Experts were asked to rank whether they agreed (1 = "strongly disagree"; 5 = "strongly agree") with the items and to suggest improvements. Hereby it was assumed that a minimal level of consensus is 50% (McKenna 1994).

7.4 Results and Discussion

7.4.1 Results of the First Round

Based on the opinions of the experts, the organic market of WBC is still in the initial phase of its development, although some countries have more advanced organic markets than the others. However, in comparison to the developed European countries, there is a vast difference in production and consumption of organic products.

At the international level, the WBC organic market started to develop during the 1990s. In WBC, there were several companies exporting such products at that time. However, back then, there was no systematic approach in most countries.

Expert 1
There was no insight into what was happening: where, by which principles, no ideas on how to develop, promote and expand the business. So, we missed all those things because we were isolated during the 1990s. But trends from developed countries influenced the Serbian organic market to start developing 10 years ago. We started to get involved in the global trends in 2000. The first Law on organic production came into force in 2000. However, the federal law from that year was not functional; it made things too formal and slow. Initiatives mostly come from the local non-governmental organizations and associations of producers.

Expert 2
Almost 14 years ago, the largest pharmaceutical company in Macedonia initiated an organic tea programme which is still running and even expanding today. That provoked smaller entrepreneurs to invest in processing facilities and now these people are running a full scale of activities in order to satisfy their own needs for quality, quantity and marketing. Thus, it could be concluded that in the past the big 'player' was a single group leader, but now there are several companies that successfully run the business by establishing cooperative relationships with the local population. Regardless of the level of achievement, the situation is far from being satisfactory and yet more needs to be done in the area of promotion and farmers' education on market demands and needs.

According to first round results, there are good preconditions for organic production in all the WBC due to the low intensity of agriculture in the past. There is a limited variety of organic products offered in terms of diversity and quantity, and producers have neither the awareness nor enough knowledge about how to produce them. The biggest problem at the moment is the fact that the organic food market is not sufficiently organised, and that the sector lacks support from the state. Experts

Table 7.3 Characteristics of the organic consumer in the WBC

Characteristics	Consumer of organic food in the WBC
Age	25–40 years of age
Gender	Mainly female
Level of education	Higher level of formal education
Revenues	Higher revenues ("well off")

state that there is a lack of efficient distribution channels in WBC. In order to develop a distribution channel of organic food, it is necessary to intensify the contacts between farmers and consumers through supermarkets and exporters of agricultural products. Consumers' motives for buying organic food are health, trend, prestige and environmental awareness. Organic consumption in WBC is becoming a part of today's lifestyle which corresponds to global trends. In addition, due to higher prices of organic food, which is often 30% more expensive than conventional food, a lot of consumers consider organic consumption to be a sign of prestige. However, with a decrease in price supplements for organic food that is expected in the future, consumption of organic food will become an even stronger part of a new lifestyle.

According to first round results, the main consumer group in WBC consists of people, mainly female, with higher revenues, higher education and of 25–40 years of age (Table 7.3). Over the next 10 years, the importance of the organic market will grow. Export of the raw organic material, import of processed organic food, the growth in number of organic farms and an increase in consumption will influence the development of the organic market in all the countries. There is a need for better promoting organic food and for informing the consumers about the organic production. Experts state tourism, agro ecotourism and public procurement (organic food in schools, hospitals, etc.) as an opportunity for further market development.

7.4.2 Results of the Second Round

The second Delphi round brought a deeper insight into the organic food market through six previously mentioned thematic sections. The results of the second round are summarised in Table 7.4 according to thematic sectors for the present as well as for the future situation with comments regarding specific country issues.

Experts stated that the National Development Strategy for organic farming is very important for the market development and will be even more important in the future. State and local governments can significantly improve the market conditions through public procurement. Governments should put special emphasis on the control system and certification. Besides motivation of the producers, teaching and training for organic farming is also important for the development of the organic sector. Practically none of the investigated countries considers meat as a market for organic products. A better situation than for meat has been identified for milk, fruits, vegetables and baby food with significant differences among countries. In the future, strong positive changes are expected. There is no market for

Table 7.4 Summary of the results of the second Delphi round

Thematic sections	Today's situation	Future predictions	Specific country issues
Government and policy impact	Importance of National Development Strategy and public procurement	Increase in importance of National Development Strategy and public procurement	Bosnia and Herzegovina and Croatia experience lack of political support and adequate incentives. Future predictions are pessimistic
Production	Motivation of producers important for further development. Financial consolidation of organic farms is necessary	Motivation of producers to convert into organic will remain a key factor for future growth of the organic sector. Financial consolidation remains a necessity	The requirement of horizontal cooperation between organic farmers and pooling of organic production has only been mentioned in Macedonia
Market	Market is increasing. Meat is not considered as a market for organic products. Clear labelling is necessary for further enhancement of the market. Distributors have higher bargaining power than producers	Market will continue to increase. Positive changes are expected for meat and other products. Clear labelling is crucial for further development. Bargaining power of distributors over producers will diminish	Organic market of Bosnia and Herzegovina is not increasing. Opinion that offer and demand of the organic market are well balanced found approval only in Croatia
Trends	Marketing of organic products is inadequate. Organic market development depends on general level of wealth. Organic sector brings new possibilities for income and labour. Ecotourism is important for development	Professional marketing for organic products is getting more important. Development of the organic sector will still depend on general level of wealth. Ecotourism will be even more significant for further development	In Macedonia it is considered that organic sector does not support economy in rural areas. Future expectations are more optimistic

Supply chain	Quality management and traceability are important for a better organisation of the organic supply chain. Distribution channels are scarce and generally underdeveloped	Increase in importance of quality management and traceability. Increase and diversity of distribution channels is expected	Supermarkets as distribution channels are considered unimportant only in Macedonia. Quality management is not important in Montenegro for a better organisation of the supply chain. In both cases future predictions are contradictory
Consumer behaviour	Main characteristic of organic consumer is high level of formal education. Health is the most important factor for buying organic food. Key promotional activities are education and information of consumers	In the future, organic consumers will no longer be characterised by higher levels of formal education. Health will remain the most important factor with increase in importance of environmental consciousness and animal welfare. Communication towards consumers will be even more important	Local agriculture, traditional agriculture, small-scale agriculture and traditional processing do not present motivation for buying organic food in Croatia and Macedonia. It is expected that this motivations increase significantly in the future

organic products in urban touristic towns, rural touristic areas, rural remote areas and rural close to urban centres. Future developments regarding these areas are supposed to lead to stronger markets. When considering trends, vertical cooperation and definition of strategy are important for the development of the organic sector. This is expected to intensify in the future. Organic farming is not a strong factor of success for the agricultural sector in all WBC. In the future, its importance for success is expected to be slightly higher. Practically none of the investigated countries agrees that distribution channels for organic products are numerous and diversified. Expectations for the future go in a strongly opposite direction. The mentioned issue will be even more important in the future. Organic consumption will become a part of a future lifestyle.

The third Delphi round was based on data derived from the second round, after repeating certain questions. We chose ten questions from the second round that met a low level of consensus. The third round questionnaire was based on the decision of the methodological group and of the leader of this project with the following aims:

(a) To reiterate important questions to get more clear answers from the experts.
(b) To test some of the questions from the second round in order to see if a higher level of consensus is possible.
(c) To get a final feedback from the WBC experts regarding future expectations. It is to be noted that some questions were reversed because of a high level of common disagreement of experts on that questions in the second round.

Table 7.5 contains a summary of key findings in the third Delphi round.

The influence of the EU Agricultural Policy will push the organic farming in WBC. The most important market variables are adequate marketing activities and a clear labelling of organic products. Concerning certain categories of organic products, the experts rank them as follows: fruits, vegetables and imported organic products, which will dominate the markets in the future. The lowest support is visible for meat. The importance of channels of distribution reached the highest level of consensus in all the examined countries because of their role in the development of the organic sector. Experts state the importance of a better organisation of the supply chain by implementing transparency, traceability and quality management. Health concerns are considered to be the most important motives for buying organic food. Experts emphasise the importance of promotional activities in education and information of consumers. With expectations of lower prices of organic foods and increased health awareness of consumers, organic consumption will become a part of the future lifestyle.

7.5 Conclusions

Organic production is now fully regulated and harmonised with the EU regulation in the WBC. These regulations will generate a better recognition of the organic food products, and this will facilitate exchanges between countries. The control of the

Table 7.5 Summary of results of the third Delphi round

Thematic sections	Future perspective	Specific country issues
Government and policy impact	Influence of EU Agricultural Policy will push the organic farming in WBC	Experts in Bosnia and Herzegovina, Croatia and Montenegro do not expect that state incentives and political support for organic sector will be adequate in the future, while in Macedonia, Serbia and Slovenia, the opposite is expected
Production	Horizontal integration between organic farmers is needed in the future	Importance of horizontal integration between organic farmers in the future is highest in Croatia and Serbia
Market	The market for organic products will be in urban areas and somewhat less in tourist areas	In Montenegro it is expected that imported organic products will dominate the organic market to a certain degree in the future
Trends	Organic farming is a success factor for agricultural sector in the future	Experts in Serbia do not consider organic farming as a success factor for agricultural sector in the future
Supply chain	Better organisation of the supply chain is necessary	Discounters are not an important distribution channel of organic food in most countries except Montenegro and Serbia
Consumer behaviour	Organic consumption will become a part of a lifestyle. Consumers will be somewhat confused by different indications for food (organic, untreated, traditional, etc.)	Consumers in Macedonia and Croatia will not be confused by different indications for food. Slovenian experts expect the most in comparison to others that prices of organic products will be too high, while none of the Macedonian experts think that

certification system, on which governments in all countries should put special emphasis, is a key success factor recognised by all experts.

The organic sector is expected to grow in the Balkan countries. However, even in Slovenia and Croatia, which are the most advanced countries in the area, the development is still slow and has to face a lot of challenges, both from the supply and demand side.

7.5.1 Government and Policy Impact

With a high level of consensus, experts consider that financial consolidation of organic farms will be necessary to enhance organic farming. As organic farms are becoming larger and larger in the Western European countries, we can suggest that a lot of investments into small organic farms of the Balkan countries will be necessary to be more competitive. However, experts' opinions vary on the important issue regarding subsidies, with strong differences between countries. More than two thirds of the respondents think that incentives will not be adequate in the future in Bosnia or Montenegro, while about half of the respondents in Croatia, Slovenia or Serbia think the same.

The stability and continuity of the national policy in the frame of a global strategy is very important for farmers as it was strongly emphasised by all the experts: a long-term engagement from governments is needed.

7.5.2 Production

It seems, according to the experts' responses, that it would be easier to produce organically in Bosnia, Macedonia and Serbia than in Slovenia or Croatia (as the former have had difficulties in the last decades to get chemical inputs in some areas the conversion would be easier). The main question is how to motivate farmers to convert. According to the experts, higher subsidies, higher product prices and the current market situation are the main motivations for farmers to convert into organic farming. To convert "industrial agriculture" into organic farming is a long-term investment and concerns the whole life of the farmer. Probably, higher skills than in conventional agriculture are needed to reach sufficient crop yields and to have profitable farms. This would have an impact on organic food price reduction, emphasised as an important restriction for the current consumption of organic food in WBC.

It is also important to mention the last decades' strong modernisation of agriculture in WBC. Farmers have been pushed to increase their yields and to turn to a modern and intensive agriculture. While huge changes in productivity are required, switching back to organic farming is difficult for farmers. Additionally, it might be important for farmers to change their attitudes and to value quality over quantity. The stakeholders have to take into account this specific psychological barrier. This will help farmers to quickly adapt to the specificities of the organic market.

Farmers' education in organic farming is crucial, and this will be probably easier to reach with younger and dynamic farmers than with older ones. They should be supported by specialised agronomists. Moreover, horizontal integration among organic farmers will be needed in the future. In addition, it is possible to suggest that cooperation with Western European producers, connecting them at a horizontal level with WBC farmers, could be very attractive and a good incentive to help the latter to convert their production.

Finally, the majority of experts consider that the interest of producers to convert into organic production in WBC will increase and that organic farming will expand, despite of the economic crisis (except in Bosnia and Herzegovina). The trend will finally be the same as in Western European countries.

7.5.3 Markets and Distribution Channels

The marketing channels have to be developed in the domestic market. Export opportunities exist and are considered as a key development factor particularly in Croatia and Serbia. Data collected during the study show that exports today are rather constituted by raw products than by processed food.

It remains important to develop the organic production and to respond to the demand in the domestic market. A majority of experts foresee that imported products will dominate the market (mainly in Slovenia and Croatia). It is also important to underline that for all experts, the main organic markets will be in urban and touristic areas.

A good marketing strategy is a pressing need because there is a risk for farmers to be forced to sell at "conventional" prices. As emphasised by all the experts, structuring of the supply chain and cooperation within marketing channels are key success factors at all stages: cooperation between producers and consumers as well as between producers and processors.

Having national products on supermarket shelves will be the result of a global project between all the stakeholders. The challenges will be the following:

(a) The quantities: producers have to be organised in order to offer the required quantities.
(b) The quality: producers have to improve their overall techniques to achieve a high level of quality for their products including value-added production (production technology, packaging, etc.).
(c) The consistency of the organic production: producers have to constantly supply markets with the requested quantity and quality.

Finally, to push supply and pull demand will be the main strategies needed in the whole area of WBC. It is important to help producers to convert their farms into organic ones and to be present in the different marketing channels. It is interesting to mention that some leader companies (e.g. in Croatia) convince farmers to convert to organic production and at the same time organise sales on the market. This supposes, as emphasised by all experts with a high level of consensus, that vertical cooperation, transparency, traceability and quality management are improved. All these measures contribute to both a better organisation of the supply chain and to guaranties regarding organic production.

In parallel, producers and all stakeholders of the sector have to communicate and to promote organic food together. Interestingly, all kinds of distribution channels (and not only supermarkets or specialised stores) are expected to distribute organic food in the Balkans. Direct selling on farms and green markets or modern ways of selling—such as via Internet or box schemes (a system of direct selling where clients subscribe to a regular delivery of farm products)—are also widely considered to be used in the future. HoReCa (Hotel, Restaurant and Catering) is not forgotten, in connection with the expected sales in touristic areas. For each channel, producers should implement a specific communication strategy, in order to get a higher added value of their production. However, experts are generally sceptic concerning the future level of promotion in the Balkans: the majority thinks that promotion will remain at a low level.

7.5.4 Consumer Demand

In the WBC, consumer awareness of healthier lifestyles and of a higher quality of food intake is expected to grow. Health has been described as the main motivation to buy organic food by all experts in all WBC. This finding is fully in line with motivation studies in Western European countries (Magnusson et al. 2003). Traditional processing as well as local and traditional agriculture are also considered as important motivations for buying organic food. This suggests that consumers in the Balkans are "natural products oriented". They are convinced that past agricultural practices used less chemical inputs. Organic consumption appears to meet both traditional values still anchored in the culture of that region and Western influences such as environmental protection and health concerns.

The majority of respondents agreed with the statement describing organic food consumers as urban, female and with a high level of education. Interestingly, elderly people are not considered to be the target group for consuming organic food. Also, in some countries, to be well off is not considered as a mandatory condition for consuming organic food. Only a minority of experts consider that organic prices will be too high in the future. Organic consumption could become a valuable choice in all social classes as it is more and more the case in Western European countries.

From the experts' point of view, education and information of consumers are key success factors. Therefore, it seems very important that organisations such as consumer organisations or public health institutes provide nutritional messages and keep explaining to consumers (starting by educated ones) what is good for their health and how they can protect the environment of their country.

7.5.5 Trends

The most important influences that have shaped the development of the organic market in the WBC are the following:

(a) Implementation of national and EU policies and regulations concerning organic production and processing.
(b) Government initiatives for the certification of organic food and incentives for producing organically have been crucial for the further development of the sector.
(c) Organic products were recognised by consumers as healthier than conventional food, and consumers assume that the organic way of production is preserving the environment.

Currently, the WBC's organic sector meets with important restrictions. However, all experts forecast that organic farming will spread in all WBCs despite difficulties. The organic farming is expected to support the economy in rural areas, preserving jobs and income. Organic production could also be connected with different types of tourism—agro-, eco- and health tourism—which is also a developing sector in the whole region.

References

Adler M, Ziglio E (eds) (1996) Gazing into the oracle: the Delphi method and its application to social policy and public health. Jessica Kingsley Publishers, London

Agence Bio (eds) (2011) L'agriculture Biologiques – Chiffres clés, Edition 2011

Armstrong JS (1978) Long-range forecasting: from crystal ball to computer. Wiley, New York

Cerjak M, Mesić Ž, Kopić M, Kovačić D, Markovina J (2010) What motivates consumers to buy organic food: comparison of Croatia, Bosnia Herzegovina, and Slovenia. J Food Prod Mark 16(3):278–292

Colom-Gorgues A (2009) The challenges of organic production and marketing in Europe and Spain: innovative marketing for the future with quality and safe food products. J Food Prod Mark 21(2/3):166–190

Critcher C, Gladstone B (1998) Utilising the Delphi technique in policy discussion: a case study of a privatised utility in Britain. Public Adm 76(3):431–449

Fowles J (1978) Handbook of futures research. Greenwood Press, Connecticut

Gatewood RD, Gatewood EJ (1983) The use of expert data in human resource planning: guidelines from strategic forecasting. Hum Resour Plann 5(1):83–94

Gordon TJ, Hayward H (1968) Initial experiments with the cross-impact matrix method of forecasting. Futures 1(2):100–116

Grundy M, Ghazi F (2009) Research priorities in haemato-oncology nursing: results of a literature review and a Delphi study. Eur J Oncol Nurs 13:235–249

Helmer O (1977) Problems in futures research: Delphi and causal cross-impact analysis. Futures 1977:17–31

Henson S (1997) Estimating the incidence of food-borne *Salmonella* and the effectiveness of alternative control measures using the Delphi method. Int J Food Microbiol 35:195–204

Hsu C-C, Sandford BA (2007) The Delphi technique: making sense of consensus. Pract Assess Res Eval 12:1–8

Jones E (1989) Reading the book of nature. University Press, Columbus

Keeney S, Hasson F, McKenna HP (2001) A critical review of the Delphi technique as a research methodology for nursing. Int J Nurs Stud 38:195–200

Linstone HA, Turoff M (eds) (1975) The Delphi method: techniques and applications. Addison-Wesley, Reading

Linstone HA, Turoff M (2002) Introduction. In: Linstone HA, Turoff M (eds) The Delphi method: techniques and applications. New Jersey Institute of Technology, Newark, pp 3–12

Loo R (2002) The Delphi method: a powerful tool for strategic management. Polic Int J Police Strateg Manag 25(4):762–769

Magnusson MK, Arvola A, Hursti UK, Aberg L, Sjoden PO (2003) Choice of organic foods related to perceived consequences for human health and to environmental friendly behaviour. Appetite 40(2):109–117

Martino JP (1983) Technology forecasting for decision making. Elsevier Science Publishing, New York

McKenna HP (1994) The Delphi technique: a worthwhile research approach for nursing? J Adv Nurs 19:1221–1225

Mullen PM (2000) When is Delphi not Delphi? Discussion paper 37, Health Services Management Centre, University of Birmingham, Birmingham

Powell C (2003) The Delphi technique: myths and realities. J Adv Nurs 41:376–382, Retrieved from: http://www.scopus.com. Accessed 10 July 2012

Rodriguez-Diaz A (2000) Globalisation and technology management in the mexican food industry. Ind Manag Data Syst 100(8–9):430–435

Sackman H (1975) Delphi critique. Expert opinion, forecasting, and group process. Lexington Books, Lexington

Sahota A (2010) The global market for organic food & drink. Weltleitmesse für Bio-Produkte, str. 1–13. (Unpublished work prepared for the Sustainable Food summit held 10–11 June 2010 in Amsterdam). Retrieved from: http://orgprints.org/18210/. Accessed 19 Aug 2011

Scapolo F, Miles I (2006) Eliciting expert's knowledge: a comparison of two methods. Technol Forecast Soc Change 73:679–704

Schaer B (2012) A glimpse on organic markets in SEE. Session at the BioFach Congress of Feb. 16, 2012. Global market overview: organic market developments in established & key emerging markets

Soon JM, Davies WP, Chadd SA, Baines RN (2012) A Delphi-based approach to developing and validating a farm food safety risk assessment tool by experts. Expert Syst Appl 39:8325–8336

Stärk KDC, Wingstrand A, Dahl J, Mřgelmose V, Lo Fo Wong DMA (2002) Differences and similarities among experts' opinions: *Salmonella enterica* dynamics in swine pre-harvest. Prev Vet Med 53:7–20

Wentholt MTA, Rowe G, König A, Marvin HJP, Frewer LJ (2009) The views of key stakeholders on an evolving food risk governance framework: results from a Delphi study. Food Policy 34:1731–1738

Wentholt MTA, Fischer ARH, Rowe G, Marvin HJP, Frewer LJ (2010) Effective identification and management of emerging food risks: results of an international Delphi survey. Food Control 21:1731–1738

Willer H, Kilcher L (eds) (2011) The world of organic agriculture. Statistics and emerging trends 2011. IFOAM, Bonn, & FiBL, Frick

Žakovska-Biemans S (2011) Polish consumer food choices and beliefs about organic food. Br Food J 113(1):122–137

Chapter 8
Utilising Conjoint Analysis: Understanding Consumer Preferences for Traditional Food

Corinne Amblard, Elise Prugnard, Georges Giraud, and Cristina Mora

Abstract The Western Balkan countries have a strong heritage of culinary tradition. In the Focus Balkans project, the conjoint analysis method has been used in order to measure the consumers' trade-offs between preferred levels of several attributes of traditional food products. By applying a conjoint analysis, researchers could gain a better understanding of the real value consumers attach to certain attributes when making purchasing decisions. The goals of the study described in this chapter therefore were to:

(a) Measure the attitudes towards traditional food products expressed by Balkan consumers.
(b) Assess the purchasing behaviour of traditional food product consumers.
(c) Identify specific segments of consumers sensitive to traditional food products.

In the scope of the study implementation, focus group discussions were held in the exploratory stage and the core activity – a conjoint analysis – was followed by an analysis of different clusters of consumers. Thereby, four specific segments of consumers who represent a different sensitivity to traditional food were identified. The information obtained can be useful for product design, strategic advertising, market segmentation and further market research questions.

C. Amblard, M.Sc. (✉) • E. Prugnard, M.Sc.
Department of Food Quality and Economy, UR CALITYSS, Clermont Université, VetAgro Sup, BP 10448, 63000 Clermont-Ferrand, France
e-mail: corinne.amblard@vetagro-sup.fr; elise.prugnard@vetagro-sup.fr

G. Giraud, M.D., Ph.D.
Department of Social Sciences and Humanities, UMR CESAER INRA-AgroSup Dijon, 26 Bd Dr Petitjean, BP 87999, 21079 DIJON cedex, France
e-mail: georges.giraud@dijon.inra.fr

C. Mora, Ph.D.
Department of Food Science, University of Parma, Via Kennedy 6, 43039 Parma, Italy
e-mail: cristina.mora@unipr.it

D. Barjolle et al. (eds.), *Food Consumer Science: Theories, Methods and Application to the Western Balkans*, DOI 10.1007/978-94-007-5946-6_8,

Keywords Consumer preferences • Post hoc segmentation • Ranking-based conjoint analysis • Traditional food • Western Balkan consumers

8.1 Introduction

In the Focus Balkans project, the conjoint analysis methodology has been used in order to measure the consumers' trade-offs between preferred levels of several attributes of traditional food products. This permitted to provide a detailed understanding of the importance of traditional food according to consumers in Western Balkan countries (WBC). According to Guerrero et al. (2009, p. 348), European consumers perceived a traditional food product as 'a product frequently consumed or associated with specific celebrations and/or seasons, normally transmitted from one generation to another, made accurately in a specific way according to the gastronomic heritage, with little or no processing/manipulation, distinguished and known because of its sensory properties and associated with a certain local area, region or country'. As underlined by the Food and Agriculture Organization (FAO) (Bernardoni et al. 2008), the preservation of traditional products could be a response to the modernisation process in transition economies. Despite the heterogeneity in terms of geography, culture, food and history, WBC might show some congruency with respect to preferences towards traditional food. By using conjoint analysis, researchers could gain a better understanding of the real value consumers attach to certain attributes when making purchasing decisions in a retail situation. This method is indeed often called upon to study the factors that influence consumers' purchasing decisions. These factors may be product attributes such as price, colour, ingredients, guarantee, environmental impact, point of sale, geographical origin and others more. Consumers typically do not have the option of buying the product that is best in every attribute, particularly when one of those attributes is price. Consumers are then forced to make trade-offs as they decide which products to purchase. This method allows consumer preferences for a product or service to be broken down into trade-offs among its individual attributes, without separating those attributes from the context in which overall consumer judgements are made.

8.2 Method

8.2.1 Origin and Purpose

Conjoint analysis has been used in research for many years (Green and Srinivasan 1978). The concept of conjoint analysis is described by Hair et al. (1998, p. 392) as follows: 'Conjoint analysis is a multivariate technique used specifically to understand how respondents develop preferences for products or services. It is based on the simple premise that consumers evaluate the value of a product or service by combining the separate amounts of value provided by each attribute'. Conjoint analysis aims at deciphering 'the composition rules used by decision makers to combine information into overall judgments' (see p. 3 in Raghavarao et al. 2011). Sudman and Blair (1998) warn that it is not a data analysis procedure. It must be regarded as a type of 'thought experiment' designed to show how various elements of products or services (price, brand, style) predict customer preferences for a product or service. Kotler (2000, p. 339) defines conjoint analysis as '…a method for deriving the utility values that consumers attach to varying levels of a product's attributes'. Churchill and Iacobucci (2002, p. 748) refer to conjoint analysis as '…conjoint measurement, which relies on the ability of respondents to make judgments about stimuli'.

This method is a popular marketing research technique. It is used in designing new products, changing or repositioning existing products, evaluating the weight of labels or brands, evaluating the effects of price on purchase intent and simulating market share. It is used for optimising product configurations, studying price elasticity of demand, simulating market response to new or modified offerings and diagnosing competitive strengths and weaknesses.

The value of conjoint analysis lies in the fact that it estimates how much each of these attributes is valued, and as Churchill and Iacobucci (2002, p. 748) state, '…the word *conjoint* ("*CONsider JOINTly*") has to do with the notion that the relative values of things considered jointly can be measured when they might not be measurable if taken one at a time'.

8.2.2 Data Collection and Processing

A ranking-based conjoint analysis has been carried out in 2010/2011 in the six WBC. In the scope of the study implementation, different activities have been planned which are the following:

- Focus group discussion
  This stage could be considered as an exploratory stage in order to provide the following conjoint analysis protocol.
- Conjoint analysis
  This activity was the core activity. Conjoint analysis measured consumers' trade-off between attributes of traditional food. This provided a detailed understanding of the importance of traditional food according to consumers in WBC.
- Analysis with a special focus on clusters
  Based on the results of the conjoint analysis, a classification method has been used in order to identify specific segments of consumers sensitive to traditional food. The classification method applied to the data (individual utilities for each level) is an Agglomerative Hierarchical Clustering (AHC) using Ward algorithm.

The sample was constituted of 1,200 respondents (200 respondents in each country to allow segmentation of the sample). In each country, the sample was split in two: 100 respondents in a rural area and 100 respondents in an urban area were interviewed. Interviewees were recruited via a filter question ('Do you consume fresh cow cheese?') in order to select only fresh cow cheese consumers in the study.

It was decided to put respondents in a purchase situation for cards ranking in order to measure purchase preferences and not the perception of traditional food. It was also agreed to work on a common product for the six WBC so that there were enough consumers in the sample and so that results could be compared between all countries. Concretely, during focus groups, different kinds of cheese were quoted as traditional products in each country. The protocol has been validated after several discussions with fresh cow cheese (Mladi Sir) as a common final choice. In order to strengthen this choice, we asked in the additional questionnaire if it was really perceived as traditional or not.

The choice of four independent attributes qualifying the fresh cow cheese was accepted: geographical origin, mean of production, price and packaging. These four attributes and their respective levels are described hereafter:

- The *geographical origin*
  3 levels: localised in the region, localised in the country and no origin specified
- The *mean of production*
  3 levels: on-farm production, small dairy production, industrial production
- The *price* (adapted to each country)
  3 levels equally spaced: lowest price, medium price, highest price of the market
- The *packaging*
  2 levels: sold loose (on desired weight) and prepacked (sealed in a plastic bag/box)

Between 54 possible profiles ($3 \times 3 \times 3 \times 2$), an orthogonal design of experiments was obtained and generated 9 profiles. In order to assess the predictive quality of the model, 2 holdouts were added to the orthogonal design. These 'holdout' cards were presented to the respondent in the same way and at the same time as the other cards. Their rank was recorded in the results but not used to calculate utilities. The characteristics of the 11 profiles are described in Table 8.1.

To make the exercise more realistic, it was required to have a picture of fresh cow cheese on each card, the same one in each country. Respondents were asked to rank the 11 cards from the most preferred one to the least preferred. During implementation, all respondents had at their disposal the same information presented in the same way to avoid any bias. After this ranking stage, respondents were invited to answer an additional questionnaire. The aim of this questionnaire was to provide explanations and complementary information on the results of the conjoint analysis task.

Table 8.1 Description of the eleven profiles kept for the cards' design

	Card	Geographical origin	Packaging	Price	Production
1	A	Localised in the country	Sold loose	High	Industrial production
2	B	Localised in the country	Prepacked	Medium	Small dairy production
3	C	No geographical origin specified	Prepacked	Low	Industrial production
4	D	Local	Sold loose	Low	Small dairy production
5	E	Localised in the country	Sold loose	Low	On-farm production
6	F	No geographical origin specified	Sold loose	Medium	On-farm production
7	G	Local	Prepacked	High	On-farm production
8	H	Local	Sold loose	Medium	Industrial production
9	I	No geographical origin specified	Sold loose	High	Small dairy production
10[a]	J	No geographical origin specified	Prepacked	High	On-farm production
11[a]	K	Local	Sold loose	Low	Industrial production

Source: Authors (Output SPSS V19.0)
[a]Holdouts

8.2.3 Needed Resources

Several interviewers were recruited and trained in each WBC. Concerning the practical implementation of the survey, feedbacks from the WBC were quite positive with generally not much trouble encountered with recruitment of willing respondents thanks to the budget allocated to reward the participation of consumers by vouchers or gifts.

The data processing was done by means of SPSS 19.0 and SPAD V7.0.

8.2.4 Benefits of the Method

Conjoint analysis has several advantages. For example, the first concern in conducting a quantitative study about traditional food was that it would not be possible to obtain concrete answers about traditional food. This difficulty comes from the fact that there is not an official definition of traditional food. Of course traditional food may be protected, for instance, food with geographical indication, but this definition is too restrictive. According to consumers, the definition of traditional food is wider, from natural products grown in the garden to cooked meals that have been cooked in the households for many decades. It may also be a typical meal of the region sold at a restaurant or even a local prepacked cheese sold in supermarkets. Using conjoint analysis permits therefore to obtain more realistic results.

Another advantage of conjoint analysis is that consumers are placed in a purchase context. It is easier for them to express their preferences in a more concrete situation instead of answering a declarative questionnaire. So rather than asking successive questions about one feature at a time, conjoint analysis measures trade-offs between several features.

8.3 Context of the Study and Objectives

WBC have a strong heritage of culinary tradition; the consumption of traditional food is strongly connected with their cultural habits. In some WBC, many products are already registered and protected in a legal frame. However, the field of traditional food is certainly larger than the products registered under a legal framework. Furthermore, it is interesting to have an overview of Western Balkan consumers' perception towards traditional food because WBC are very heterogeneous in terms of geography, culture, food and history. These different items may influence the nature of traditional food in each country. In the scope of the project Focus Balkans, the objective of this task was to better understand consumers' attitudes, expectations and behaviour towards traditional food in six WBC. The goals were to:

(d) Measure the attitudes towards traditional food products expressed by Balkan consumers
(e) Assess the purchasing behaviour of WBC consumers for traditional food products
(f) Identify specific segments of consumers sensitive to traditional food products

8.4 Results

8.4.1 Sample Description

The additional questionnaire administered at the end of the conjoint analysis task permitted to gather socio-demographic data from the 1,200 respondents interviewed. The sample is equally divided between the six countries. The sample of respondents is described according to socio-demographic information in Table 8.2.

8.4.2 Perception of Fresh Cow Cheese

Regarding Table 8.3 about answers given to the question 'In your opinion, is fresh cow cheese a traditional product?', it seems that fresh cow cheese has been a really good choice as a relevant example of traditional food in all WBC studied. The main reason for the minority of consumers who do not find fresh cow cheese traditional is that fresh cow cheese can be found everywhere. It means that for these consumers, traditional food is strongly linked to a locality and should not be exported or produced in other countries or maybe regions.

Table 8.2 Part1: sample description (%)

		BiH[a] (%)	Croatia (%)	Maced. (%)	Monte. (%)	Serbia (%)	Sloven. (%)	Rural (%)	Urban (%)	Total (%)
Grew up in a rural area	Yes	49.0	65.5	60.5	43.0	38.5	86.5	68.8	45.5	57.2
	No	51.0	34.5	39.5	57.0	61.5	13.5	31.2	54.5	42.8
Gender	Male	25.0	33.0	51.5	47.5	44.5	45.5	40.3	42.0	41.2
	Female	75.0	67.0	48.5	52.0	55.5	54.5	59.7	57.8	58.8
Age cluster	25 and –	15.0	28.0	20.5	17.0	25.5	26.5	18.7	25.5	22.1
	26–35	31.0	20.5	13.5	36.0	20.5	21.5	24.7	23.0	23.8
	36–45	20.5	13.5	15.5	25.5	16.5	21.5	17.3	20.3	18.8
	46–55	23.5	24.0	30.0	14.0	21.5	13.5	22.7	19.5	21.1
	56 and +	10.0	14.0	20.5	7.5	16.0	17.0	16.7	11.7	14.2
Education level	Unfinished elementary school	0.0	1.0	0.5	0.5	1.0	0.5	0.8	0.3	0.6
	Finished elementary school	1.5	5.0	16.0	4.5	7.0	14.0	8.2	7.8	8.0
	Finished secondary school	64.5	62.0	58.5	53.5	60.0	54.5	56.2	61.5	58.8
	Finished college	14.5	14.0	9.5	20.0	12.5	10.0	16.0	10.8	13.4
	Finished faculty	19.5	18.0	15.5	21.0	19.5	21.0	18.8	19.3	19.1
Current occupation	Unemployed (currently)	9.5	10.5	15.0	13.5	19.0	4.0	14.7	9.2	11.9
	Student/trainee	4.5	22.0	17.0	2.0	22.0	20.5	10.3	19.0	14.7
	Pensioner	6.5	10.5	17.0	4.0	13.0	15.0	10.8	11.2	11.0
	Unskilled and skilled worker	44.0	8.0	8.0	0.5	5.5	2.0	12.2	10.5	11.3
	Farmer/fisherman	1.0	1.0	7.5	0.0	1.0	3.5	3.3	1.3	2.3
	Storekeeper/trader/craftsman	5.0	3.5	15.5	1.0	5.5	9.0	6.3	6.8	6.6

(continued)

Table 8.2 (continued)

		BiH[a] (%)	Croatia (%)	Maced. (%)	Monte. (%)	Serbia (%)	Sloven. (%)	Rural (%)	Urban (%)	Total (%)
	Employee, intermediate profession	7.0	30.0	16.5	68.0	18.0	29.0	31.3	24.8	28.1
	Manager/executive	3.5	7.5	2.5	5.5	6.5	6.0	4.3	6.2	5.3
	Other	19.0	7.0	1.0	5.5	9.5	11.0	6.7	11.0	8.8
Household net income (HNI) in the last month[b]	No income	1.0	3.0	1.0	0.5	4.0	2.0	2.0	1.8	1.9
	≤ Minimum HNI	51.5	29.0	44.5	55.0	27.0	29.5	46.8	32.0	39.4
	> Min. HNI and ≤ average HNI	28.5	28.0	39.5	31.0	37.5	29.0	30.3	34.2	32.3
	> Average HNI	7.0	22.5	14.0	7.0	21.5	15.0	8.8	20.2	14.5
	Not disclosed	12.0	17.5	1.0	6.5	10.0	24.5	12.0	11.8	11.9
Number of household members	1	1.5	4.5	3.0	1.5	3.5	1.0	3.3	1.7	2.5
	2	18.0	9.0	12.5	16.5	12.5	14.5	16.2	11.5	13.8
	3, 4	54.0	57.0	55.5	53.0	66.5	44.5	53.0	57.2	55.1
	5 and +	26.5	29.5	29.0	29.0	17.5	40.0	27.5	29.7	28.6

[a] *BiH* Bosnia and Herzegovina
[b] Adapted per country

Table 8.3 Proportion of respondents according to the categorisation of fresh cow cheese as a traditional product or not

	'In your opinion, is fresh cow cheese a traditional product?'						
	BiH (%)	Croatia (%)	Macedonia (%)	Montenegro (%)	Serbia (%)	Slovenia (%)	Total (%)
Yes	94.5	82.0	86.0	92.5	78.0	66.5	83.3
No	2.0	2.5	8.5	2.0	6.0	17.0	6.3
Do not know	3.5	15.5	5.5	5.5	16.0	16.5	10.4
Total	100.0	100.0	100.0	100.0	100.0	100.0	100.0

8.4.3 Purchasing Behaviour

- Frequency of purchase
  A quarter of Western Balkan respondents purchase fresh cow cheese more than once a week and nearly 50% purchase it one to four times a month. It is a traditional product with a regular consumption pattern in WBC.
- Place of purchase
  The most quoted place of purchase for fresh cow cheese is the supermarket followed by open markets, friends or family, farm and at other places. Restaurants are not a place of purchase for this product. Rural respondents more often buy fresh cow cheese from friends or family and less at markets and supermarkets than average, and the tendency is reversed for urban consumers.
- Own production or friends/family production of fresh cow cheese
  Less than a quarter of the respondents sometimes make their own fresh cow cheese. About half of the consumers received home-made fresh cow cheese from friends or family. There is a significant expected difference between rural and urban areas, with more own production in rural areas (32.2% sometimes or frequently) and less in urban areas (16.7%) and with donation of fresh cow cheese from friends or family also more common in rural areas (60.0%) than in urban (42.7%).
- Type of packaging
  The most quoted packaging for usual fresh cow cheese purchases is on desired weight (sold loose). Second, it is bought already sealed in a plastic box and eventually already sealed in a plastic bag. Nevertheless, as a single response was requested, summing the last two categories show that the prepacked product is more often chosen than the one sold loose.

8.4.4 Consumption Behaviour

8.4.4.1 Consumption Frequency

As can be seen in Fig. 8.1, more than half of the sample consumes fresh cow cheese more than once a week and more than 80% consume it once a month or more often.

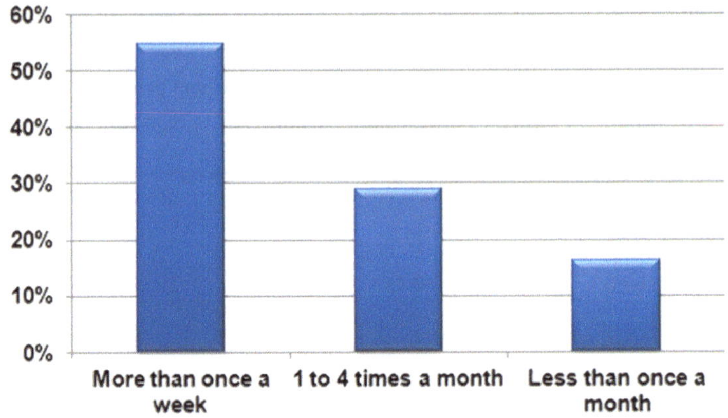

Fig. 8.1 Distribution of respondents according to consumption frequency of fresh cow cheese (*Source*: Authors)

The main barriers to its consumption quoted are in the first place high prices. This is followed by the difficulty to find it and finally its high fat content or the preference for other cheese.

8.4.5 Consumer Preferences

8.4.5.1 Cards Ranking

A first summary of data gathered during the conjoint analysis ranking task is obtained by calculating frequencies for each card according to the rank considered. Mean ranks permit to identify most preferred and least preferred cards. We can also calculate cumulative frequencies for ranks 1st + 2nd and 10th + 11th to better summarise information. It gives a first general but still interesting glimpse on data, deciphering which card had been most/least preferred. These results can be seen in Table 8.4. The cards to which they referred are presented in Fig. 8.2.

Referring to cumulative frequencies (rank 1st + 2nd and 10th + 11th), the most preferred card is G. Its combination of levels is a local origin, an on-farm production, a high price and a prepacked presentation. On the other hand, the card most often ranked 10th or 11th is card C which presents a fresh cow cheese sold without specified origin, produced industrially, sold at a low price but still prepacked.

The reasons standing behind preferences of cards are not clearly identified at this stage as the impact and importance of each attribute is not calculated yet. As a matter of fact, respondents had to make trade-offs as their ideal combination of levels was not necessarily available among the cards displayed. Some attributes were therefore considered as more essential than others. In the following paragraph, still at an aggregated level, the importance value in decision-making of each attribute will be calculated to determine the different degrees of influence, so as the utilities of levels

Table 8.4 First descriptive statistics about cards ranking

Card letter	Mean rank	% of resp. rank 1	% of resp. rank 1 or 2	% of resp. rank 11	% of resp. rank 10 or 11
Card A	6.75	6.25	12.67	17.33	27.25
Card B	4.97	11.25	24.50	2.67	5.42
Card C	6.65	9.67	15.67	16.75	28.08
Card D	5.30	11.17	23.67	5.42	12.33
Card E	5.29	14.25	25.00	6.08	12.33
Card F	5.90	7.33	16.00	5.58	14.92
Card G	5.04	17.75	30.42	6.08	12.67
Card H	6.38	5.00	12.58	6.00	18.58
Card I	6.67	5.42	11.08	10.92	22.67
Card J	6.06	7.83	17.83	9.42	20.08
Card K	7.00	4.08	10.58	13.75	25.67

Fig. 8.2 Pictures of corresponding cards (*Source*: Authors)

to compare which are the most preferred or less preferred ones. A further segmentation will describe the different types of behaviours and characterise the consumers from each cluster identified.

8.4.5.2 Attribute Importance

Table 8.5 displays the importance values for attributes and the utility value of each level. Results were obtained thanks to SPSS conjoint procedure allowing to analyse data gathered as sequences about card preferences.

Table 8.5 Importance of each attribute and utilities of their levels

Factors	Modalities	Averaged importance (%)	Utilities
Production	On-farm	31.8	0.37
	Small dairy		0.19
	Industrial		–0.55
Price	Low	28.7	0.11
	Medium		0.12
	High		–0.23
Origin	Local	25.4	0.24
	Country		0.19
	No origin		–0.43
Packaging	Sold loose	14.1	–0.21
	Prepacked		0.21

Source: Authors (Output SPSS V19.0)

Utilities are only comparable when belonging to the same attribute. A positive utility conveys that a level is more appreciated, whereas a negative utility conveys that it is less appreciated. For instance, some levels such as no origin and industrial are the least preferred levels of their respective attributes (origin and production).

Regarding the importance values, the level of production is the most important purchase criterion, followed by the price and the geographical origin. Type of packaging is by far the least relevant attribute for consumers.

The ideal combination (ideal card) for the average Balkan consumer would be fresh cow cheese produced on-farm, sold at a medium price, with a local specified origin and sold prepacked, whereas the least preferred combination would be fresh cow cheese produced industrially, sold at a high price, without any geographical origin specified and sold loose. Globally, there is a preference for small production of fresh cow cheese with a specific localisation suggested by a locality or the country. The consumers would not buy this product with a high price and there is a slight preference for prepacked cheese instead of cheese sold loose.

8.4.5.3 General Fit of the Model

A predictive validity can be estimated by calculating a utility for each card per each respondent and by calculating the correlation between those results and the actual ranks given. This correlation is given by Pearson's R or Kendall's tau in Table 8.6. All the correlation coefficients being very close to one, it shows that correlations are significant. This means that the model has a good predictive quality. Moreover, even if prediction on cards, which created the design, is generally overconfident, the cross validation making prediction for holdouts from the model created from the nine cards confirms the model quality.

Table 8.6 Correlations between observed and estimated preferences

	Value	Sig.
Pearson's R	0.998	0.000
Kendall's tau	1.000	0.000
Kendall's tau for holdouts	1.000	

Source: Authors (Output SPSS V19.0)

8.4.6 Consumer Segmentation

The analysis at the aggregate level of results is not sufficient to really comprehend consumer behaviour. There is a need to study this behaviour at an individual level and identify different segments of behaviours (post hoc segmentation). In post hoc segmentation schemes (Green 1977), estimation of the conjoint analysis occurs at the individual level, and subsequent level part-worths are then clustered to form market segments (DeSarbo and DeSarbo 2003). Respondents belonging to one segment will present similar preferences for the different attribute levels of the product. Some of these segments may be interesting to analyse in depth due to their size or the sensitivity of their members towards a concept. We can study which kind of product and which combination of attribute levels would be attractive for a particular segment and try to match consumers' desires.

The clusters are built from the table of individual utilities. Then, the segments can be characterised and explained by supplementary data from an additional questionnaire. Importance values help us to identify which attributes are playing the main role for each cluster. Four clusters are selected. The final stable clusters obtained are described in the following paragraph and the mean utilities and importance values per cluster are displayed in Table 8.7.

*Cluster*1: Importance of origin (weak attractiveness of products without any geographical origin specified): 216 respondents (18% of total sample)

Cluster 1 constitutes a segment of consumers for whom geographical origin of fresh cow cheese is a main concern. They clearly prefer local fresh cow cheese and are less fond of this product when no origin at all is specified. After having taken into account the origin of the product, their attention goes to the means of production. They still prefer on-farm production and less prefer industrial production. However, this weaker attractiveness of industrial products is not as distinctive as seen on average in the total sample; that is not what makes the distinction of this segment. Price and packaging have little importance and no really specific behaviour appears considering these attributes. Moreover, in this class, importance of the attribute 'origin' is significantly higher than on average.

In this cluster, no country is significantly overrepresented. However, Macedonia is underrepresented (10.65 against 16.67% in the total sample), and this may be linked to the fact that fresh cow cheese is apparently not as often produced locally as in the other countries (a bit less traditional). Cluster 1 is characterised by the fact

Table 8.7 Importance and utility values detailed per cluster

	Cluster	Cl. 1	Cl. 2	Cl. 3	Cl. 4	Total
Utilities						
Geographical origin	Local	1.92	−0.11	0.02	−0.23	0.24
	No origin	−2.18	0.00	−0.19	−0.01	−0.43
	Country	0.26	0.12	0.17	0.24	0.19
Packaging	Sold loose	−0.11	0.05	−0.25	−0.47	−0.21
	Prepacked	0.11	−0.05	0.25	0.47	0.21
Price	Low price	0.06	0.10	1.98	−1.01	0.11
	Medium price	0.04	0.12	0.40	−0.01	0.12
	High price	−0.10	−0.22	−2.38	1.02	−0.23
Production	Small dairy production	−0.02	0.60	0.17	−0.09	0.19
	On-farm production	0.37	1.72	−0.05	−0.66	0.37
	Industrial production	−0.35	−2.32	−0.12	0.74	−0.55
Importance values						
Geographical origin		49.07	20.24	16.87	22.28	25.41
Packaging		10.93	13.21	13.70	16.99	14.10
Price		18.66	17.72	49.91	31.68	28.72
Production		21.34	48.82	19.53	29.05	31.77

that there are slightly more consumers declaring that they never consume fresh cow cheese with the family on festive occasions (12.0 instead of 7.6% on average) and more consumers declaring they did not consume it when they were a child or cannot remember it. Households from cluster 1 are a bit larger than on average.

Cluster2: Importance of the means of production, weak attractiveness of industrial products: 364 respondents (30% of total sample)

Cluster 2 constitutes a segment of consumers who place the means of production as their main concern for fresh cow cheese purchases. They prefer on-farm production, they also like fresh cow cheese from small dairies, and they clearly prefer less this product when produced industrially. By decreasing importance in decision-making, we find far behind geographical origin, then price and finally packaging, all of them with no particular stressed preferences from this segment. In this cluster, attention paid to production is significantly higher than on average. On the other hand, the importance of origin and price is lower than the average importance.

In this cluster, Croatia (26.92 against 16.67% in the total sample) and Slovenia (21.43 against 16.67% in the total sample) are significantly overrepresented, whereas Macedonia is strongly underrepresented (2.75 instead of 16.67% in the total sample). Interviewees are less keen on supermarkets for fresh cow cheese purchases than the average of the total sample and favour friends, family, on-farm or own production. There are proportionally fewer respondents who think that food with geographical indication (GI) is traditional. Moreover, consideration about the necessity of higher hygiene is slightly less important than on average, whereas the country of origin is a bit more important for them than on average.

*Cluster*3: Weak attractiveness of high prices: 238 respondents (20% of total sample)

In this third cluster, the main concern of consumers is the price of fresh cow cheese. We find indeed a linear relation for this attribute, with the weakest preference for high prices, a moderate acceptance of medium prices and a clear preference for low prices. All other criteria seem not to have any weight in decision-making for these consumers. Despite the fact that industrial products or products with no specified origin are least preferred regarding their respective attributes, the gap with the other levels is not really stressed. It means that these consumers may favour price even if it can mean a lack of quality.

In this cluster, Montenegro (26.47 against 16.67%), BiH (22.69%) and Serbia (22.69%) are overrepresented, whereas Macedonia (5.04%) and Croatia (6.30%) are underrepresented. Consumers from cluster 3 buy their fresh cow cheese and cheese more often on green or open markets than the tendency of all consumers. This seems to convey that fresh cow cheese is more affordable on open markets. These respondents are more often the main cooks of their household (48%). They or their family/friends make their own fresh cow cheese less often and they receive less often home-made cheese from friends or family than the average. High price is more often given as a reason of no willingness to increase consumption. There is a slightly higher agreement on the statement about the link between taste of the food and the region of origin than on average. Regarding the socio-demographic variables, households are a bit smaller than the average size and the proportion of consumers having finished faculty is lower.

*Cluster*4: Preference for industrial, prepacked products with high prices: 382 respondents (32% of total sample)

This cluster is an unexpected one. Consumers of this segment gave main importance to two attributes: first to the price and then to the means of production. Strangely, they favour fresh cow cheese when sold at a high price but also when sold industrially. The third attribute considered is origin and finally the packaging which is more important here than in other clusters. Products with no origin specified are a bit more preferred than on average (mean utility around 0 compared to the negative one in the total sample) and prepacked fresh cow cheese is clearly more preferred than on average.

In this cluster, Macedonia is overrepresented (40.58 against 16.67% in the total sample), whereas BiH (9.16%), Slovenia (9.42%) and Croatia (11.26%) are underrepresented. The consumers of cluster 4 are less often the main buyers and cooks of their households with regard to the average proportion. The consumption as child and during festive occasions is more important. Their preference for industrial fresh cow cheese at a high price is coherent with the fact that they buy less frequently at farms or from friends and family. They buy fresh cow cheese more often in supermarkets than the average and own production is more common in this cluster for cheese and fresh cow cheese in particular. The desire to consume fresh cow cheese more often is inferior compared to other clusters. We also studied the income for this segment because of their willingness to pay higher prices, and we just notice a more important proportion of respondents who do not quote the income of the household. They more often consider food with geographical indication as traditional and significant characterisation shows they give more importance to hygienic consideration (stricter for market/farmers and more confident towards own production) than on average but, even if still high, have less consideration than the average respondent for geographical origin of food. Maybe they buy industrial fresh cow cheese in order to meet their requirements for hygiene and thus are willing to pay more for this attribute.

8.5 Discussion

By trying to decipher consumer behaviour towards traditional food, different types of consumers giving more or less importance to components considered as traditional have been highlighted (Fig. 8.3). To summarise what has been highlighted in the six segmentations done by country and keeping in mind what we learn from the global classification, four types of behaviour stand out:

- *Consumers sensitive towards levels identified as rather traditional*: Giving weakest preference to industrial products and/or to products without specification of geographical origin.
- *Consumers looking for convenient purchase*: Attracted by prepacked food (the characterisation of this cluster can explain this behaviour as it includes in proportion more young people from 26 to 35 years old who are not strongly deprived, so their lifestyle may lead them to choose convenience and easiness over other attributes). A cluster only highlighted for BiH consumers.

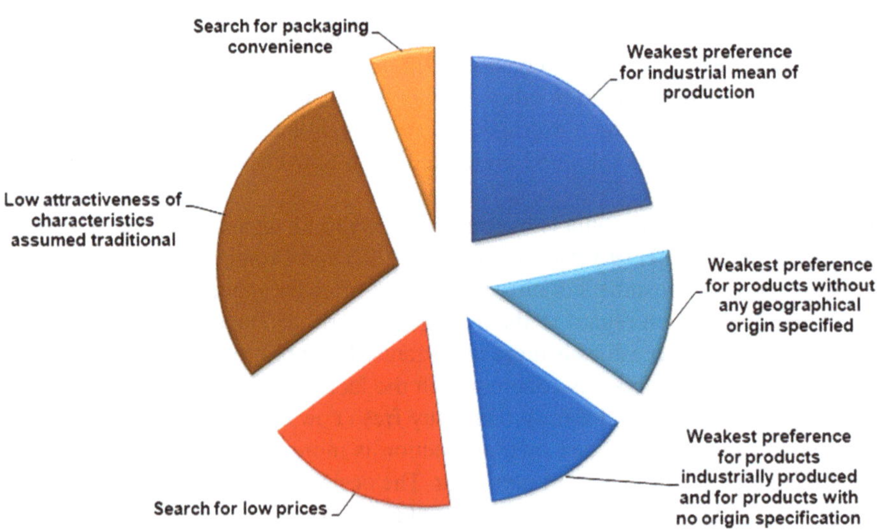

Fig. 8.3 Relative weight summarising the behaviours identified through the segmentations by country (*Source*: Authors)

- *Consumers favouring low prices*: A behaviour that conveys mostly a situation of poverty.
- *Consumers giving lower importance to traditional levels*: More strongly trusting industrial products and willing to pay higher prices for the additional confidence from these products (not really trusting the hygiene practices of small-scale food processing).

Conjoint analysis, coupled with a declarative questionnaire, is an interesting method, but it needs a long preparation: identification of the product chosen, choice of attributes and levels, experimental design, creation of cards and other steps more. It is also necessary to interview respondents face to face in a comfortable situation. This is why this method is not very convenient for interviews in the street. A constraint to take into account is also that cognitive capacities of respondents are weak. The number of attributes to consider for getting a realistic description of the product concept is large. However, conjoint analysis requires a limited number of attributes in order not to propose consumers to rank too many cards, which will reduce the reliability of results. Lastly, using such a method for a cross-cultural study requires taking into account cultural differences between countries. For example, there are language differences; therefore, it may be difficult to find a common product allowing the comparison of results between countries.

Finally, the information obtained from this method can be applied to different market research questions. It can be useful for product design, strategic advertising, market share, cost-benefits analysis and market segmentation. Often used in market research applications, conjoint analysis can also be interesting to measure people's perceptions or judgements. The information gained from a conjoint study could also be used for the development of a theoretical model towards understanding consumer apparel purchasing decisions (North and de Vos 2002).

8.6 Conclusions

In conclusion, this survey about consumers' attitudes, expectations and behaviours towards traditional food is innovative in the Balkans and highlights some interesting results. Conjoint analysis, also called trade-off measurement, was used to determine the main attributes of a traditional product expected by the consumers and to measure the weight and attractiveness of targeted elements in consumer perception. However, the method has some limits. The results are applied on fresh cow cheese and not on traditional food in general. Fresh cow cheese is traditional in all WBC though. This may highlight a specific behaviour towards cheese, so it will be interesting to conduct the survey on another category of products such as meat. Moreover, even if a lot of similarities exist among the six WBC studied, different consumer behaviours towards traditional products have been highlighted between the countries.

These results could be strengthened by the other quantitative survey held in the scope of the Focus Balkans project which measures notably attitudes towards food in general, the consumption and purchase behaviour towards traditional dishes and the perception of geographical indication. We could also compare these results with those obtained in the European project TRUEFOOD focused on traditional food perception in European countries and using conjoint analysis method (Naes et al. 2010). It would also be interesting to do a supplementary analysis with EuroFIR results, the European Food Information Resource Network of Excellence, where some definitions of traditional food based on reviews of national and European regulations have been developed.

Lessons Learnt and Utility for Marketing Management

The implementation of the study on consumers' preferences for traditional food in WBC required to explore the diversity of traditional food in these countries and showed the bond of the population with such food. Taking into account this diversity and the cultural differences between the countries is necessary for the implementation of the conjoint analysis, especially in the scope of a cross-cultural analysis. Then, results are interesting and may be used by marketing managers to promote traditional food in different countries. Coupled with cluster analysis, managers may identify different market segments for their products and, this way, better understand the market preferences. Knowing which value consumers place on certain features of a product or service is useful for market decisions. Conjoint analysis may also be used by managers to design new products or adapt those existing in order to comply with consumers' needs.

References

Bernardoni P, Esteve M, Paus M, Reymann R (2008) Case studies on quality products linked to geographical origin in the Balkans. FAO, Agridea, Lindau

Churchill G, Iacobucci D (2002) Marketing research, methodological foundations, 8th edn. Harcourt Publishing, London

DeSarbo W, DeSarbo C (2003) A generalized normative segmentation methodology employing conjoint analysis. In: Gustafsson A, Herrmann A, Huber F (eds) Conjoint measurement: methods and applications. Springer, Berlin, pp 473–504

Green PE (1977) A new approach to market segmentation. Bus Horiz 20:61–73

Green PE, Srinivasan V (1978) Conjoint analysis in consumer research: issues and outlook. J Mark 55:103–123

Guerrero L, Guardia MD, Xicola J, Verbeke W, Vanhonacker F, Zakowska-Biemans S, Sajdakowska M, Sulmont-Rossé C, Issanchou S, Contel M, Scalvedi ML, Signe Granli B, Hersleth M (2009) Consumer-driven definition of traditional food products and innovation in traditional foods. A qualitative cross-cultural study. Appetite 52:345–354

Hair JF, Anderson RE, Tatham RL, Black WC (1998) Multivariate data analysis, 5th edn. Prentice-Hall International, Upper Saddle River

Kotler P (2000) Marketing management. The millennium edition. Prentice-Hall, London

Naes T, Lengard V, Bolling JS, Hersleth M (2010) Alternative methods for combining design variables and consumer preference with information about attitudes and demographics in conjoint analysis. Food Qual Prefer 21:368–378

North E, de Vos R (2002) The use of conjoint analysis to determine consumer buying preferences: a literature review. J Fam Ecol Consum Sci 30:32–39

Raghavarao D, Wiley JB, Chitturi P (2011) Choice-based conjoint analysis: models and designs. CRC Press, Boca Raton

Sudman S, Blair E (1998) Marketing research. McGraw Hill International Editions, Boston

Chapter 9
Conclusions

Dominique Barjolle, Pascal Bernardoni, Jasna Milošević Đorđević, Goran Zivkov, Dragana Tar, and Boban Mugoša

Abstract Understanding consumer behaviour plays a major role in strengthening the competitiveness of the food industry and in improving the well-being of Western Balkan citizens. Whilst their governments are seeking to best manage the transition from planned to market-based economies, the lack of data and applicable models in food consumer science hinder these efforts significantly. Three reasons call for a more developed food consumer science in the region: the rapidly evolving markets that globalise domestic trade; the diverging preferences of consumers related to changing living standards requiring new marketing strategies; and the importance of information on food availability and dietary patterns for the development of food policies. This book serves as a dissemination tool for the results of the European research project Focus Balkans and as a textbook for prospective scientists. It summarises the main findings of the related case studies and explains the methods

D. Barjolle, Ph.D. (✉)
Institute for Environmental Decisions (IED), ETH Zürich,
Sonneggstrasse 33, SOL C2, CH-8092 Zürich, Switzerland
e-mail: barjolle@ethz.ch

P. Bernardoni, M.Sc.
SEEDEV in Serbia, Koce Kapetana 47, 11000 Belgrade, Serbia
e-mail: pascal.bernardoni@seedev.org

J.M. Đorđević, Ph.D. (✉)
Ipsos Strategic Marketing, Gavrila Principa 8, 11000 Belgrade, Serbia
e-mail: jasna.milosevic@ipsos.com

G. Zivkov • D. Tar, M.A.
SEEDEV in Serbia, Koce Kapetana 47, 11000 Belgrade, Serbia
e-mail: goran.zivkov@seedev.org; dragana.tar@seedev.org

B. Mugoša, Ph.D.
Institute of Public Health-Montenegro,
Džona Džeksona bb, 81000 Podgorica, Montenegro
e-mail: boban.mugosa@ijzcg.me

D. Barjolle et al. (eds.), *Food Consumer Science: Theories, Methods and Application to the Western Balkans*, DOI 10.1007/978-94-007-5946-6_9,

applied. In this chapter, conclusions concerning four major evolving markets (health claim products, fruits, organic products and traditional food) and findings on food choices of Western Balkan consumers are presented. It is concluded that institutional capacity building is urgently needed. The development of consumer rights and of public health policies must be supported by a Europe-wide network of scientists and public health officials. This book aims at contributing to these important developments.

Keywords Food choice motives • Qualitative approach • Quantitative survey • Food choices • Institutional capacity building • Consumer rights • Public health policies

9.1 Food Consumption in the Western Balkan Countries: Concluding Remarks

In the Western Balkan countries (WBC), the lack of data and appropriate models of food consumption hinders understanding of important changes that are currently occurring in food markets. Indeed, these countries are seeking to best manage the transition to more market-based and 'Europeanized' economies. The requirement to develop food consumer science derives from several reasons.

First, markets are rapidly evolving: The changing nature of domestic markets leads to an increasing presence of the major international and domestic supermarkets. Whilst the market is open and more accessible to private investments, the supermarkets are gaining market shares. This tendency plays an important role in forming requirements at the farm and local food processing level. In particular, local actors must comply with increasingly demanding global requirements, for example, the sanitary regulations.

Second, consumers' preferences vary and diverge substantially according to evolving living standards. These differences are the result of the evolution of purchasing power, but this differs depending on the region and the social belonging of the consumer. An appreciation of these new regional and social frameworks should be integrated into the marketing strategies of producers. New opportunities appear: For example, the demand for novelties (e.g. 'health claim products'), organic products and traditional foods is growing both on domestic and export markets. Market and consumer research is able to support producers in their commercial efforts to cope with changing consumer requirements.

Third, information on food availability and dietary patterns in the WBC is essential for the development of food policies designed to ensure sufficient food supply and improved human health and well-being.

As a consequence, the need for an improvement in market participation is obvious: The Western Balkan food industry (farmers, processors, producers and retailers) requires a better understanding of consumer demand and consumption trends. This presupposes the collection of the necessary information and its presentation in a form accessible for food supply chain actors. It was a primary objective of the Focus Balkans European research programme and of this book, as a tool for dissemination, to provide research results. In order to explore in-depth certain aspects of food markets, the research undertaken in the Focus Balkans project was focused on four major evolving markets: health claim products, fruits, organic products and traditional food. In some parts of the region, difficulties remain in calculating useful statistics. For instance, censuses are not updated in all countries, and the generally patchy and low-quality data on production and retailing hinder understanding of general consumption patterns.

Nevertheless, combining expert opinions and consumers' perceptions using a qualitative approach and subsequently conducting a representative quantitative survey, the Focus Balkans project builds a good basis for better understanding food motives and choices, consumer behaviour and future trends on certain food markets in the region.

9.1.1 Food Choices Motives

The most important food choice motives in the six Western Balkan countries under review are: 'sensory appeal', 'health and natural content', 'price' and 'purchase convenience' of the food. Differences between the six countries are commented on in Chap. 3. This ranking of food choice motives is similar to the ones in other European countries, that is, Great Britain (Steptoe et al. 1995; Pollard et al. 1998), Belgium, Italy (Eertmans et al. 2006), Russia (Honkanen and Frewer 2009), Finland (Lindeman and Väänänen 2000) and Greece (Fotopoulos et al. 2009) (see Table 3.3). As in other studies, 'ethical concern' and 'familiarity' were found to be the least important motives for dietary choice in the WBC.

Exploratory factor analyses yielded eight factors instead of nine as in the initial UK study. The main differences were 'natural content' and 'health' loaded onto one factor, whilst 'convenience' split into two, which we labelled 'purchase convenience' and 'preparation convenience'. A single 'health factor' containing items relating to health benefits and natural characteristics (absence of artificial ingredients or additives) was also obtained in research for Belgium and Italy (Eertmans et al. 2006), suggesting that food consumers view both the safety and the nutritional value of the food to be essential for its healthiness. Russian food consumers (Honkanen and Frewer 2009) also viewed preparation convenience to be different from purchase convenience (availability), with higher importance attributed to the latter motive.

Confirmatory factor analysis did not support the initial nine-factor structure proposed by Steptoe et al. (1995). Although the normative study is more than 15 years old, there are recent data suggesting that for some samples the nine-factor structure still holds true (Bomba et al. 2011), so our findings cannot solely be attributed to changes in the global market environment but also to the specific characteristics of food consumers in the WBC.

The research conducted on a sample most similar to this one (Fotopoulos et al. 2009) is particularly relevant. Indeed, the Balkan Peninsula is one geographical region, considered by some authors as presenting consistent similarities from a cultural point of view (Prévélakis 1996). The results presented by Fotopoulos et al. (2009) also suggested reducing the number of factors from the original nine to five or even four as well as the collapsing of the second order motivations in order to assess the more fundamental motivations behind food choices. There are some similarities between the WBC and Greece. This research study used an existing validated instrument to capture food motivations. Although developing a more appropriate questionnaire tailored to the region, as suggested by Fotopoulos et al. (2009), was not an objective of the project, this will be relevant for further research.

9.1.2 Health Claim Products

The market for products with health claims is one of the fastest growing segments of the European and worldwide food industries. The take-off of the market is already evident in Slovenia and Croatia and upcoming in the rest of the WBC (Serbia, Montenegro, Bosnia and Herzegovina (BiH) and Former Yugoslav Republic of Macedonia (FYROM)). Young, educated, wealthy and urban women are the main consumers. The health claims are related to nutritional labelling. Thus, the analysis of the WBC market, which is rather underdeveloped, is necessarily based on the observation of products with nutrition and health claims (N&H claims). The WBC consumers tend to prefer domestic products.

According to the quantitative analysis, health claim products are perceived as 'healthy but also as tasty and pleasant, suggesting that this product category has an overall positive but undifferentiated image' (Žeželj et al. 2012). The main barriers to the development of the market are consumer related. Expert interviews (see presentation in Chap. 5) reveal that eating habits were the most important problem quoted by interviewees, followed by consumers' lack of awareness and knowledge. Bearing in mind the rising importance of the markets for health claim products, some retailers in Slovenia have developed their own private labels for healthy food, thus giving less space to private companies using their brand names. Processors in WBC are highly dissatisfied with national policies for nutrition. Namely, in most of the countries, such policies have not been developed (Serbia, Montenegro, FYROM, BiH), or, if existing, they are not properly implemented (Slovenia).

Supplying products with health claims on the market can be useful for the image of brands: consumers may perceive this as innovative and socially responsible. It is, however, a complex process, as the health claims have to be scientifically proved and communicated properly to consumers. Based on expert interviews, the best strategy identified is to choose and combine several health effects that are the most important ones for the national health (e.g. cardiovascular diseases and diabetes) and work on their promotion together with public health institutions. The development of the market is best linked with the implementation of effective public policy.

Proactive public policy can promote the concept of healthy lifestyles and healthy eating in the region. At the same time, the regulation of the market increases the trust of consumers and supports efficiently the efforts of communication: scientifically sustained information related both to relation of diet and health and of diet and disease may be seen as reliable by the consumers.

9.1.3 Fruits

Western Balkan countries have a climate favourable for fruit growing. The consumption of fruits in the region presents many similarities due to common cultural habits and traditions. The food culture in the WBC is close to the Mediterranean diet for the regions located at the coast and close to the European food culture for eastern Croatia and Serbia. In the WBC, fruit consumption strongly decreased during the 1990s, but during the most recent years, the consumption of fresh and processed fruits has increased. Statistical evidence suggests that the consumption of fruits in general is lower than in European Union (EU) countries, although the availability of traditional fruits is high combined with low prices. However, the reliability of statistics is questionable, mainly because non-marketed output is not recorded despite the importance of home-grown fruits being very high.

Based on complementary expert interviews not presented in this book, the research made clear that the market suffers, on the one hand, from inadequate coordination of production, quality and contracts with trade and processors and, on the other hand, of the increasing influence of European retailers and their worldwide sourcing strategies to obtain low prices, high-quality and assortment standards.

Market strategies based on consumer segmentation will strengthen efforts to increase fruit consumption. Complementary health policy measures like public awareness campaigns would support increasing fruit intake. This would be particularly useful for urban and young consumers, who are adopting Western European lifestyles whilst losing the possibility to eat home-grown fruits.

9.1.4 Organic Products

Compared to developed European markets and the other Western markets in the organic food sector, the WBC markets are immature and underdeveloped. Committed market actors, some associations of organic farmers, stakeholders in rural development and environmental protection, a few consumer associations and policymakers act nevertheless proactively in favour of the expansion of the organic sector.

In the last 20 years, whilst the farming intensity in the region mostly could be characterised as ‘low’, it was not certified as official organic production. Since the year 2000, market actors and their associations but also foreign market actors, donor agencies and policymakers have tried to stimulate the 'official' organic sector regarding:

- Official regulation: In most WBC, national laws provide a framework based on the model of EU regulations, for the production and marketing of organic products.
- Certification: In all WBC, domestic or international certification bodies guarantee the respect of the domestic law and/or of international standards.

As a consequence, this clearer institutional framework has triggered production growth in all countries. For the time being, the collection of wild products still dominates sales. Furthermore, plant production is more significant than animal production. Supply chains are generally short, with much direct selling on the domestic market. The best-developed supply chains, with intensive and high-quality processing, are directed to export markets.

Consumers in the WBC show little but increasing interest in buying organic food products. The barriers are nevertheless high: there is still no clear image of what means organic. Research shows that consumers in the WBC typically confuse certified organic products with the ones sold on green markets and traditional products, believing that they are all organic, which is obviously not the case. Private sector commitment to the development of the organic market, with public support, should better inform consumers about the guidelines that are respected by organic producers and about the significance and importance of certification.

9.1.5 Traditional Food

Consumers have a strong culinary heritage in the Western Balkans. In some WBC, many products are already registered and protected according to specific legislation. However, traditional food is not only limited to registered products under a legal framework. In a context of economic transition, the existing national regulations tend to comply more and more with EU regulations for traditional products. The main objectives of this EU regulation are to maintain a diversified agricultural production, to encourage rural development of less-favoured areas, to give clear information to consumers regarding the origin of products and, eventually, to harmonise the legal protection framework for geographical indications. The EU framework also intends to prevent unfair competition between producers and enhance the public credibility of Protected Designation of Origin (PDO), Protected Geographical Indication (PGI) and Traditional Speciality Guaranteed (TSG) schemes. The Focus Balkans quantitative survey reveals that 80 % of WBC citizens think that traditional products should be protected, and consider geographical origin as an important purchase criterion for traditional foods.

However, as the concept of traditional products is wider than the PDO, PDI and TSG schemes, some other concepts such as 'on farm', 'regional' or 'local' products exist and meet consumer demand. The research underlines a lack of confidence about hygiene practices especially for small-scale production. Consumers expect that the regulation of traditional products will enhance the sustainability and the diversity of food. A segment of consumers expresses a strong willingness to buy high-quality traditional food products, even if prices are slightly higher than for conventional products. In addition, according to focus groups conducted as part of the Focus Balkans project, the negative aspect of traditional products is related to health concerns because traditional food is often considered as too fatty or salty.

At the consumer level, a general lack of knowledge about traditional food products persists. There is no clear definition and geographical indications are not well known by consumers. There may be also a confusion between organic and origin labels. Public awareness campaigns undertaken by the private sector and consumer associations with public support should provide better information to consumers about food labelling and official voluntary standards like PDO or PGI.

9.2 Institutional Capacity Building

Consumer behaviour in the WBC has changed over the recent years. Consumers' buying habits are shifting away from frequenting green markets and gardening for self-consumption to supermarkets and other retail chains. This trend represents a challenge for all the institutions, which are part of the food consumer knowledge system. They need to adapt their strategies by taking into consideration changes in consumer behaviour. This should encompass multidisciplinary research on the impacts of such changes and inclusion of related topics in educational curricula, as well as dissemination of information and awareness campaigns conducted by governments and public health organisations.

These changes are part of a wider process as the WBC shift to more market-based economies that are integrated into European and global markets. One of the most significant trends resulting from this process is the growing importance of supermarkets and hypermarkets as the main format for food shopping. Shifts in shopping habits may have negative effects on health, for increased consumption of non-traditional, energy-dense processed foods high in salt, fat and sugar (Baillie 2008).

Hawkes (2005) examines the role of foreign direct investment (FDI) in changing patterns of food consumption, focusing on highly processed foods. The effect of FDI in food processing, services and retail is to make more highly processed foods available to more people. Hence, FDI has made it possible to lower prices, open up new purchasing channels, optimise the effectiveness of marketing and advertising, and increase sales of highly processed food (Baillie 2008). FDI would therefore be an appropriate entry point to implement a range of public health policies to 'redirect' shifts in nutritional intake.

Hawkes (2008) analyses the dietary implications of supermarket development: how supermarkets make decisions – namely, the location of stores, the foods they sell, the prices they charge, the promotion strategies they use and the nutrition-related activities (healthy production lines) they implement – influence food accessibility, availability, prices and desirability. The findings are both positive – increased choice of food for more people – and negative – reduced ability of marginalised groups to access high-quality food and encouragement of the consumption of energy-dense, nutrient-poor, highly processed food. Overall supermarkets encourage consumers to eat more of whatever food. This is also confirmed by a study conducted in Guatemala (Aswaf 2007). The latter details how supermarkets in developing countries concentrate on processed, dry and packaged foods because they enable economies of scale and have long shelf lives. Most of these food items tend to contain a disproportionately high amount of added fat, sugar and salt, which are identified as potential risk factors for obesity and non-communicable chronic diseases.

Most of the WBC communication between consumers, as the end users, and other entities in food consumer science is one way. At present, state administrations and other entities of food consumer science are not paying sufficient attention to the needs and interests of the end users.

This will have to change with the development of food consumer science, as all WBC will be, sooner or later, part of the EU, and the demand and opinions of the end consumers will have to be taken into account.

9.2.1 Development of Consumer Rights

As far as consumer rights are concerned, consumer awareness and the effective implementation of the regulations about food safety and traceability should be improved.

Consumer associations with strong linkages to their bases should be further strengthened. This is, for three main reasons, particularly true as far as food-related issues are concerned: Food is an important element of everyday life, food consumption is strongly interrelated with human health and the concentration within the food industry and retail sector taking place worldwide requires strong contact persons on the side of consumers. Food production is not totally disconnected from the consumer, but consumers should be taken into account and supported by other actors of food consumer science much more, in particular governmental and state institutions.

9.2.2 Public Health Policies

Public health is a major concern in all WBC. As the transition process is still going on in Serbia, BiH, Montenegro and FYROM, there is an urgent need to make significant progress in ensuring a better general framework for food consumption. Action needs to be taken at several levels, to implement better food safety provisions or, for example, consumer rights regarding food labelling. In the framework of the Focus-Balkans programme, it was unfortunately not possible to address all the topics related to public health policy. From the beginning of the programme, some specific sectors were chosen to address this issue in a concrete manner.

Effective food and health policies should underpin the development of the market for health claim products. This will involve aligning the agro-food supply chain with healthier diets, whilst key regulative aspects of nutrition labelling have to be implemented. Government interventions should create a framework for positive industry action. Future research should be focused on the institutions that can help in better aligning the agro-food supply chain with the development and adoption of healthy diets. Institutional reforms should be targeted toward the design of the most effective tools to help public health systems to tackle more efficiently the growing problems of food-related diseases. Public health officials cannot leave this issue to chance. Extensive public programmes are required to support and facilitate the adequate consumption of food that would lead to healthier food choices. Scientists should undertake further research on specific problems such as obesity and pay special attention to vulnerable consumer groups.

9.3 Research Needs

For all the four studied markets (fruits, products with health claims, organic products and traditional food), there is a need for future research on food consumption, food and health policies, and market trends. The recommendation is to develop a systematic collection of reliable data on food consumption so as to increase scientific-based knowledge in three main areas.

The first area of research relates to consumers: food intake, food behaviour, consumers' beliefs, expectations, preferences, motives and attitudes. The main characteristics of food intake and diets are highly related to food culture and habits. Therefore, consumers in the WBC behave different from those in Western Europe. Even if the Western Balkans is a region where a strong common cultural heritage leads to similarities in the food behaviour, local differences are noticeable. Food and health policies have not yet had sufficient impact to enhance the general level of well-being of the population. Hence, it is necessary to reinforce research, particularly in these countries, both in terms of market development and strengthening public policy.

The second research topic concerns communication and information policies and tools (efficiency, targeting, content, uptake, etc.). There is a general need to improve the level and nature of communication between stakeholders within the food system, especially with consumers. Research should play a major role in developing innovative means for communication.

The third area is market research: obtaining a detailed understanding of the structure, functioning, rules, organisations, standards and level of investment on specific food markets. The research should provide qualitative and quantitative data about the trends of each market to better evaluate future opportunities and threats. The various stakeholders require these data to complement consumer research in order to elaborate appropriate strategies.

9.3.1 *Networking for Food Consumer Science in Europe*

The development of an extended European network on food consumer science would increase the quality of research in this field. This infrastructure is required to improve the real impact of research.

Food consumer science is currently a rather poorly defined field drawing on a broad range of scientific disciplines. There is therefore a need to develop new knowledge and working methods in this field. The Balkan network has been created but needs to be strengthened and expanded. Thus, a common European network, including the WBC, should be created and involve all the concerned disciplines. This is the only way to better promote new knowledge and to strengthen consumer science in Europe.

Understanding consumer behaviour and preferences plays a major role in strengthening the competitiveness of the food industry as well as in improving the health and well-being of European and Western Balkan citizens. The development of a critical mass (scientists, scientific disciplines, facilities, etc.), facilitation and promotion of data exchange, joint activities (protocols, metrics, collaborative studies) and comparability of research actions in the field of consumer science related to food is urgently required. We hope that this book will provide the necessary scientific knowledge to support this capacity building, especially by motivating future scholarly work in this highly important area.

References

Aswaf A (2007) Supermarket purchases and the dietary patterns of households in Guatemala. IFPRI discussion paper 00696, Washington, D.C.

Baillie K (2008) Health implications of transition from a planned to a free-market economy – an overview. Obes Rev 9(suppl 1):146–150

Bomba J, Modrzejewska R, Beauvale A (2011) Factor structure of symptoms in the Krakow Depression Inventory KID IO "B1" among 15-year-olds. Psychiatria Pol XLV(2):245–259

Eertmans A, Victoir A, Notelaers G, Vansant G, Van den Bergh O (2006) The food choice questionnaire: factorial invariant over western urban populations? Food Qual Prefer 17:344–352

Fotopoulos C, Krystallis A, Vassallo M, Pagiaslis A (2009) Food Choice Questionnaire (FCQ) revisited. Suggestions for the development of an enhanced general food motivation model. Appetite 52:199–208

Hawkes C (2005) The role of foreign direct investment in the nutrition transition. Public Health Nutr 8(4):357–365

Hawkes C (2008) Dietary implications of supermarket development: a global perspective. Dev Policy Rev 26(6):657–692

Honkanen P, Frewer L (2009) Russian consumers' motives for food choice. Appetite 52:363–371

Lindeman M, Väänänen M (2000) Measurement of ethical food choice models. Appetite 34:55–59

Pollard T, Steptoe A, Wardle J (1998) Motives underlying healthy eating: using the food choice questionnaire to explain variation in dietary intake. J Biosoc Sci 30:165–179

Prévélakis G (1996) Les Balkans: cultures et géopolitique. Collection Géographie d'aujourd'hui. Nathan, Paris

Steptoe A, Pollard T, Wardle J (1995) Development of a measure of the motives underlying the selection of food: the food choice questionnaire. Appetite 25:267–284

Žeželj I, Milošević J, Stojanović Z, Ognjanov G (2012) Motivational and informational basis of attitude toward food with health claims. Appetite 59(3):960–967

Authors Bios

Corinne Amblard, M.Sc., engineer Corinne Amblard is a research scientist at VetAgroSup with expertise in consumer surveys. She has a Master's degree in marketing. She has been working at VetAgroSup since 1998 on different EU research projects. Her field of expertise lies in consumer surveys, marketing and economy.

Pascal Bernardoni, M.Sc., graduated as a Master in Agro-Economics at the Federal Institute of Technology (ETH) Zürich and worked in Western Africa and Latin America. Since 1999, he has been engaged in the Balkans and in the Caucasus with specific emphasis on agriculture and rural development. Since 2006, Pascal Bernardoni has worked on various projects related to rural development with a specific emphasis on the development of geographic indication systems in Bosnia, Serbia and Croatia.

Ružica Butigan, M.B.A., is a teaching and research assistant in the Department of Marketing. She holds courses in marketing, marketing strategy and SMEs marketing. She is involved in organic food research in Croatia.

Miljan Cvetković, Ph.D., is an assistant professor and vice dean. He graduated as a M.A. in Fruit Science and worked for 2 years as general manager in the Institute for Viticulture and Enology. Since 2002, he has been working on fruit projects for the United Nations Food and Agriculture Organization in Serbia. Since 2004, Miljan Cvetković has been teaching assistant in the Department for Fruit, Agriculture Faculty of the University of Banja Luka.

Georges Giraud, M.D., Ph.D., is a professor and as senior researcher in food consumer science since 1993. He has written over 200 publications including several peer review processes in the fields of agricultural marketing, consumer science, food economics, geographical indications, rural development and retail studies. Georges Giraud was partner and coordinator of EU-funded projects.

Cristina Mora, Ph.D., has been involved as senior researcher and project leader for Italian and European research projects since 2000. She is an associate professor at the University of Parma. During her research activities, she worked on topics

D. Barjolle et al. (eds.), *Food Consumer Science: Theories, Methods and Application to the Western Balkans*, DOI 10.1007/978-94-007-5946-6,

about food marketing with special attention to food safety standards, to quality products and to consumers' willingness to pay for safer food.

Boban Mugoša, Ph.D., is an associate professor and is a specialist of epidemiology. He holds a Master's degree in epidemiology at the Medical Faculty of Belgrade. In 2009, he wrote his Ph.D. thesis in epidemiology at the Medical Faculty in Kragujevac. He is author and co-author of many books and guidelines. Since 2004, he has been working as the director of the Institute of Public Health of Montenegro.

Mitchell Ness, Ph.D., is a lecturer in marketing at Newcastle University Business School and was previously a lecturer in marketing in the Department of Agriculture, Food and Rural Development in Newcastle. He is responsible for the delivery of courses in analytical techniques for marketing and direct and digital marketing strategy. His research interests are in consumer behaviour and consumer attitudes to food quality issues. He has been involved in EU projects on products of geographical indication and quality and low-input foods. His current research activity has focused upon modelling consumer intentions towards repeat purchase of online grocery shopping in the UK and modelling the determinants of customer loyalty in the Chinese restaurant sector.

Galjina Ognjanov, Ph.D., is associate professor of marketing at the Faculty of Economics, University of Belgrade. Her research interests include integrated marketing communication, advertising research and food consumption in emerging economies. She has published three textbooks and a number of papers in scientific journals and contributed to scientific monographs published both in Serbian and in English. Since 2011, she has been serving as European Marketing Academy (EMAC) national representative for Serbia.

Jure Pohar, Ph.D., is a professor and has more than 15 years of professional experience in marketing and management. In addition to his academic carrier at the University of Ljubljana, he has held different posts in marketing-advertising companies. He was, as a member of the board of directors, responsible for marketing and R&D at the largest Slovene dairy company. He served as a national coordinator at the Slovene Ministry for science and technology and as the representative of Europe in CGIAR (Consultative Group on International Agricultural Research).

Elise Prugnard, M.Sc., engineer Elise Prugnard is research scientist at VetAgro Sup. She has a Master's degree in data analysis and graduated in statistics. Her field of experience is data extraction and analysis, statistics applied to marketing and consumer behaviour.

Natasa Renko, Ph.D., is a full professor in the Department of Marketing. She gives courses in marketing, marketing strategy, SMEs marketing and decision-making in marketing. She is author and co-author of more than 100 papers. She specialised in food marketing research. Natasa Renko is project leader of several food projects in Croatia and in the United States of America (collaborating with the United States Department of Education).

Sanda Renko, Ph.D., is an associate professor at the Trade Department and is involved in several scientific projects such as the project of the development of organic food distribution channels and consumer protection in the Republic of Croatia and the project of the development of sale and distribution systems in Croatian enterprises. In 2006, she was awarded the "Mijo Mirković" award for the best scientific paper. Today, she is the editor in chief of the *International Journal of Sales, Retailing and Marketing*.

Burkhard Schaer, Ph.D., is an agricultural engineer and agronomist and has written a Ph.D. on organic markets in Germany. Burkhard Schaer is a codirector and co-founded Ecozept in 2000. He conducted more than 180 projects on food marketing for private enterprises, research bodies and government agencies in different countries. He is implied in European research projects. His teaching activities comprise food marketing (guest lecturer) at several universities.

Siet J. Sijtsema, Ph.D., is a scientific researcher and her key qualification is consumer research on food being involved in qualitative research projects. She pays attention to the cognitive aspects as well as to the affective aspects like feelings and emotions. She is an expert on transforming consumer perception into food product characteristics in a consumer-oriented product design.

Igor Spiroski, M.D., M.P.H., is medical doctor and holds a Master's degree in public health. He works as a researcher in the Department of Physiology and Monitoring of Nutrition. His primary professional occupation is health risk assessment related to nutrition for different population groups of interest. He also works in safety assessment of functional foods. Igor is teaching assistant at the Faculty of Medicine in Skopje for the subjects of nutrition and environmental health. He has worked in several projects funded by United Nations agencies and the European Commission.

Dragana Tar, M.A., holds a Master's degree in development studies and a combined experience of work in environmental field, organisational change and rural development in the Balkans and Caucasus. Currently, she is mostly involved in the study and protection of traditional products, working as a trainer and facilitator.

Ante Vuletić, M.A., is a teaching and research assistant in the Department of Marketing. He gives courses in marketing, marketing strategy, decision-making in marketing and public relations. He is involved in organic food research in Croatia.

John White, Ph.D., is an associate professor and has acted as a coordinator for two EU INTAS projects on the restructuring of food supply chains in the former Soviet Union. His main research areas are supply chain management, especially in the agri-food industry of Central and Eastern Europe, and brand management. He teaches marketing communications and marketing theory at the University of Plymouth.

Martine Zaouche-Laniau is a graduate agro-food engineer (Ecole Nationale Supérieure des Industries Alimentaires), Master of Psychology (Université Paris V Sorbonne). She has joined Gestion Engineering Marketing (GEM) in 1980. She is in charge of market studies and supply chain analysis.

Iris Žeželj, Ph.D., has been lecturer at the Belgrade University – Faculty of Philosophy, Department of Psychology since 2003. She is a member of International Society of Political Psychology. Her fields of interest are attitude change, attitude-behaviour relation studies, media and communications and social sciences research methodology.

Karin L. Zimmermann, B.Sc., R.M., is senior researcher and project manager and his expertise lies in strategic trends and strategies based on consumer information. Karin Zimmermann has been involved in quantitative and qualitative market research, consumer and chain research and in agricultural environment (green) economics. She is leader of a research group on strategic marketing and consumer research. In the FP7 EU project FOCUS BALKANS, she is the administrative coordinator for the Netherlands and involved in consumer and market research.

Goran Zivkov Between 2004 and 2007, Goran Zivkov has worked as an Assistant Minister and Minister of Agriculture. Previously, Goran Zivkov worked for the UN FAO in Serbia. He is currently working on agriculture and rural development projects in the Western Balkan and Caucasus region.

Editors Bios

Dominique Barjolle, Ph.D., is senior researcher and lecturer at IED ETHZ. She is interested in food consumer science, quality of food and food standards. She has particular skills in the transaction cost theory in the frame of food supply chain analysis. Notably, her work on food quality linked to origin and innovations in food supply chains is recognised through numerous scientific publications. She works as an expert for the Food and Agriculture Organization of the United Nations (FAO), United Nations Industrial Development Organization (UNIDO) and the European Commission.

She participated actively in more than ten European projects and is the coordinator of the FP7 FOCUS BALKANS research programme.

Since May 2012, she is deputy director of the Research Institute of Organic Agriculture in Switzerland and, in parallel, reads lectures at the Federal Institute of Technology Switzerland (ETH).

Matthew Gorton, Ph.D., is a reader and has contributed to various projects on the restructuring of the agri-food sector in Central and Eastern Europe. This has included work on agricultural water management in the Western Balkans and rural development policy in the New Member States of the European Union. He has undertaken research on regional foods in the United Kingdom, financed by the Department for Environment, Food and Rural Affairs (DEFRA).

Jasna Milošević Đorđević, Ph.D., works as a consultant at Ipsos Strategic Marketing, as a lecturer of social psychology at the Faculty of Media and Communication at the Singidunum University and as a researcher in the Institute for Political Studies. Her Master's and Ph.D. theses were related to voting behaviour and identity. She has published 1 book and over 30 scientific articles in national and international journals. She has worked on various projects related to political public opinion and media with a specific emphasis on the development of the region.

Žaklina Stojanović, Ph.D., is an associate professor of Agricultural Economics at the University of Belgrade – Faculty of Economics, Serbia. She has held this position since 2008. She is currently a researcher at the Center of Scientific Research at

D. Barjolle et al. (eds.), *Food Consumer Science: Theories, Methods and Application to the Western Balkans*, DOI 10.1007/978-94-007-5946-6,

the Faculty of Economics, University of Belgrade (NICEF). She has participated in several national and international projects concerning the food sector. Her work also includes the fields of rural and regional development. She is the author of 4 books and over 80 articles in the areas of agribusiness, food market and agricultural and rural policy.

Index

D. Barjolle et al. (eds.), *Food Consumer Science: Theories, Methods and Application to the Western Balkans*, DOI 10.1007/978-94-007-5946-6,

FSC
www.fsc.org
MIX
Papier aus verantwortungsvollen Quellen
Paper from responsible sources
FSC® C105338

If you have any concerns about our products,
you can contact us on
ProductSafety@springernature.com

In case Publisher is established outside the EU,
the EU authorized representative is:
Springer Nature Customer Service Center GmbH
Europaplatz 3, 69115 Heidelberg, Germany

Printed by Libri Plureos GmbH
in Hamburg, Germany